Centre number
Candidate number
Surname and initials

CW01080374

General Certificate of Secondary Education

Chemistry
Higher Tier
Exam 1 Paper 1

Time: one and a half hours

Instructions to candidates

Write your name, centre number and candidate number in the boxes at the top of this page.

Answer ALL questions in the spaces provided on the question paper.

Show all stages in any calculations and state the units.
You may use a calculator.

Include diagrams in your answers where this may be helpful.

Information for candidates

The number of marks available is given in brackets **[2]** at the end of each question or part question.

The marks allocated and the spaces provided for your answers are a good indication of the length of answer required.

Where you see this icon you will be awarded marks for the quality of written communication in your answers.
This means, for example, that you should:

- write in sentences
- use correct spelling, punctuation and grammar
- use correct scientific terms.

For Examiner's use only	
1	
2	
3	
4	
5	
6	
7	
8	
9	
Total	

1 Crude oil is separated into fractions.
The diagram shows equipment used to carry out this process.

Leave blank

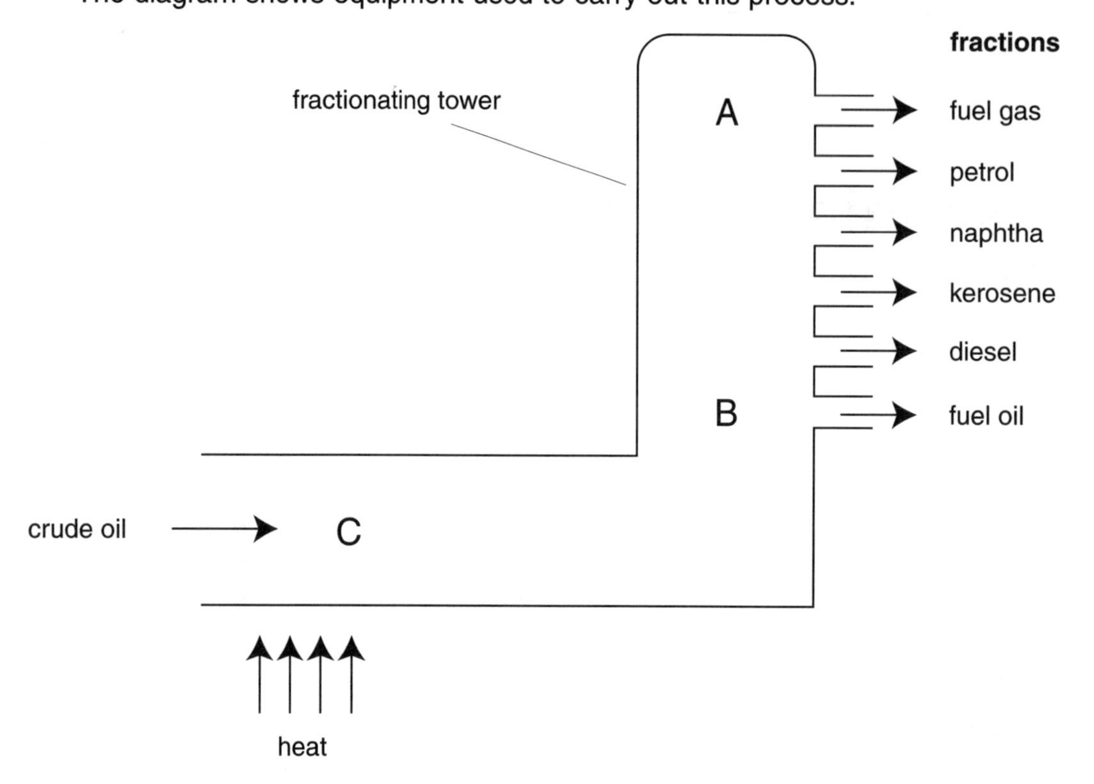

(a) Look on the diagram at the places labelled A, B and C.
At which place is the crude oil at the lowest temperature?

.... **[1]**

(b) Crude oil contains hydrocarbon molecules of different sizes.
Which of the fractions labelled on the diagram contains the largest hydrocarbon molecules?

.... **[1]**

(c) The fuel gas fraction includes the hydrocarbon ethane, C_2H_6.

(i) Draw a diagram to show the structural (displayed) formula of ethane.

[2]

(ii) Name the group of hydrocarbons which includes ethane.

.... **[1]**

(d) When ethane is passed over a hot catalyst, the molecule breaks down to form hydrogen and an alkene.

Leave blank

(i) Complete this equation for the reaction.

$C_2H_6 \rightarrow H_2 +$ **[1]**

(ii) What is the name of this alkene?

.... **[1]**

(iii) Describe a test you could carry out on this alkene to distinguish it from ethane.

test ....

result .... **[2]**

(e) Poly(chloroethene) is a polymer made by joining together molecules of the monomer chloroethene.

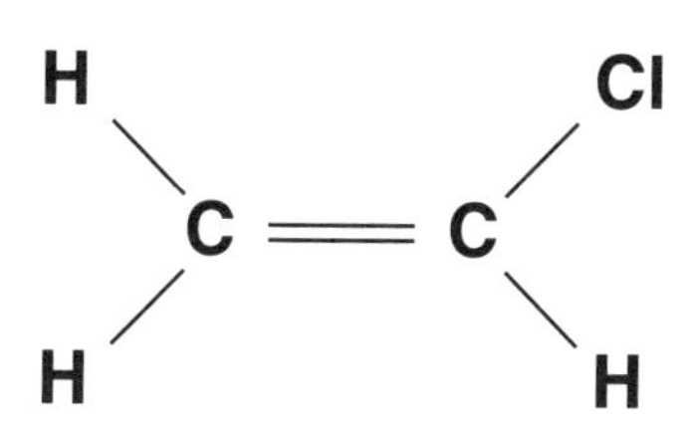

(i) Draw a diagram to show the structural (graphical) formula of poly(chloroethene).

[2]

(ii) Poly(chloroethene) is also known as poly(vinylchloride) or PVC.
State one use of PVC and give the property of the polymer which makes it suitable for this use.

use ....

property .... **[2]**

(Total 13 marks)

[turn over

Leave blank

2 The table shows the composition of dry, unpolluted atmosphere today, and a suggestion of the composition of the atmosphere 3 billion years ago.

atmosphere today	
gas	**approx percentage**
oxygen	21
nitrogen	78
carbon dioxide	0.03
ammonia	0
water	variable

atmosphere 3 billion years ago	
gas	**approx percentage**
oxygen	0
nitrogen	9
carbon dioxide	74
ammonia	7
water	variable

(a) (i) When the Earth's atmosphere was first formed about 4 billion years ago it contained hydrogen and helium.
Why were these gases not present in the Earth's atmosphere 3 billion years ago?

....

.... **[2]**

(ii) Why could animals and plants not exist in the atmosphere of 3 billion years ago?

.... **[1]**

(b) Describe how the changes in the Earth's atmosphere over the past billion years occurred.

You will be given credit for spelling, punctuation and grammar.

....

....

....

....

....

.... **[5+1]**

(Total 9 marks)

Leave blank

3 The table shows the boiling points of elements in the second row of the Periodic Table.

Group	1	2	3	4	5	6	7	0
symbol of element	Li	Be	B	C	N	O	F	Ne
boiling point in °C	1331	2487	3927	4827	–196	–183	–188	–246

(a) Compare the boiling points of the elements Li and Be with those of the elements N and O.

(i) Describe the difference in the state of these elements at room temperature.

....

.... **[1]**

(ii) Explain why there is such a large difference in the boiling points of the two pairs of elements.
Use ideas of bonding between particles in your answer.

....

....

....

.... **[4]**

(b) The element in this period with the highest boiling point is carbon.
The boiling point given is for carbon in the form of diamond.
Explain why carbon in diamond has such a high boiling point.

....

.... **[2]**

(c) The elements in the next period of the Periodic Table are:

Na Mg Al Si P S Cl Ar.

(i) Which of these elements has the highest boiling point?

.... **[1]**

(ii) Explain your choice in answer to (i).

....

.... **[2]**

Leave blank

(d) The table gives boiling points of the elements in Group 0.

symbol of element	boiling point in °C
He	–269
Ne	–246
Ar	–186
Kr	–152
Xe	–108

(i) Describe how the boiling points of these elements change down the Group.

.... **[1]**

(ii) Explain this trend in boiling points.

....

.... **[2]**

(Total 13 marks)

4 This equation shows the reaction between aluminium and iron(III) oxide commonly known as the thermit reaction.

$$2Al + Fe_2O_3 \rightarrow Al_2O_3 + 2Fe$$

Fine powders of the two reactants are first mixed together.
The reaction is then started by lighting a 'wick' of magnesium ribbon.
Molten iron is produced during the reaction. This can be used to weld together pieces of steel, for example railway tracks.

(a) Iron has a melting point of 1535°C.
Explain how **molten** iron is produced in this reaction.

....

.... **[2]**

(b) Suggest why the reaction does not begin until the magnesium ribbon is lit.

....

.... **[2]**

Leave blank

(c) (i) If aluminium powder is replaced by copper powder, no iron is produced. Explain why.

....

.... **[2]**

(ii) Suggest a metal which could successfully replace aluminium in the thermit reaction.

.... **[1]**

(iii) The thermit reaction is known as a redox reaction. What is meant by the term redox?

.... **[1]**

(d) Why are the two reactants used in the form of fine powders?

....

.... **[2]**

(e) In a thermit reaction 500 g of each powder are mixed. Calculate the maximum mass of molten iron which could be produced in the reaction.

Mass of iron = ... g **[3]**

(Total 13 marks)

[turn over

5 Copper(II) carbonate reacts with hydrochloric acid as shown in the equation.

Leave blank

$$CuCO_3 + 2HCl \rightarrow CuCl_2 + CO_2 + H_2O$$

In an experiment lumps of copper(II) carbonate were added to some hydrochloric acid and the mass of the mixture measured at intervals of time. The loss in mass during this time is shown in the table.

time in s	50	100	150	200	250	300
mass lost in g	0.20	0.35	0.41	0.44	0.45	0.45

(a) Plot the mass lost against time on the grid.
Draw the best fit curve.

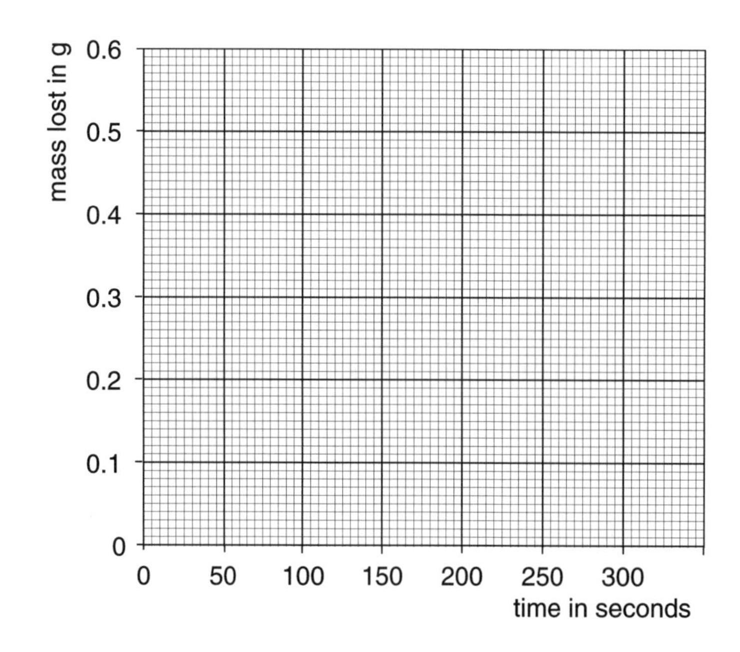

[3]

(b) Explain why the mixture lost mass as the reaction took place.

....

.... **[1]**

Leave blank

(c) Draw apparatus which could be used to carry out this experiment.

[3]

(d) The experiment was repeated using powdered copper(II) carbonate. The volume and concentration of hydrochloric acid, and the mass of copper(II) carbonate were kept the same.

(i) Sketch on the grid the graph you would expect from this second experiment. **[2]**

(ii) Explain why this graph is different from that obtained using lumps of copper(II) carbonate.

....

....

....

.... **[3]**

(Total 12 marks)

6 The diagram shows apparatus set up to study the electrolysis of a solution of sodium chloride in water.

Leave blank

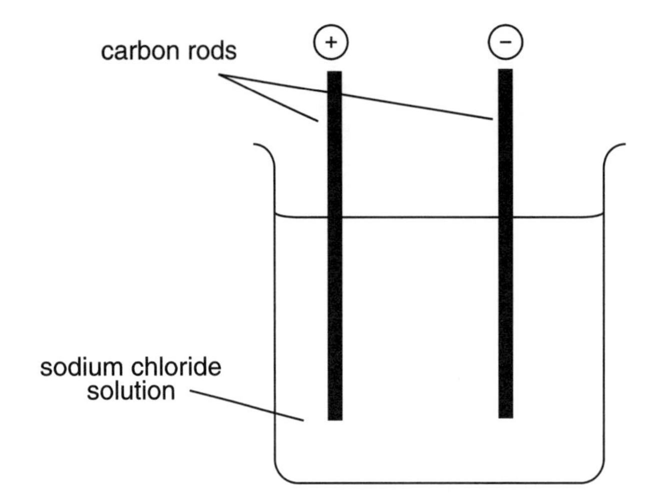

(a) A green gas was seen to bubble up from the positive electrode. This gas bleached moist blue litmus paper to white.

(i) Explain these observations.

....

....

.... **[2]**

(ii) Construct an ionic equation for the reaction at the positive electrode.

.... → .... **[2]**

(b) Before the electrolysis took place, a few drops of the indicator phenolphthalein were added to the salt solution.
During the electrolysis a colourless gas bubbled up and a pink colour appeared in the solution around the negative electrode.

(i) Name the gas given off at the negative electrode.

.... **[1]**

(ii) Explain the appearance of the pink colour.

Leave blank

....

....

....

.... **[4]**

(c) A number of useful products are obtained from the electrolysis of brine. Name two of these products, and state a use for each.

1 product ....

use ....

2 product ....

use .... **[2]**

(Total 11 marks)

7 The diagram outlines the process used to make nitric acid from ammonia and hydrogen.

Leave blank

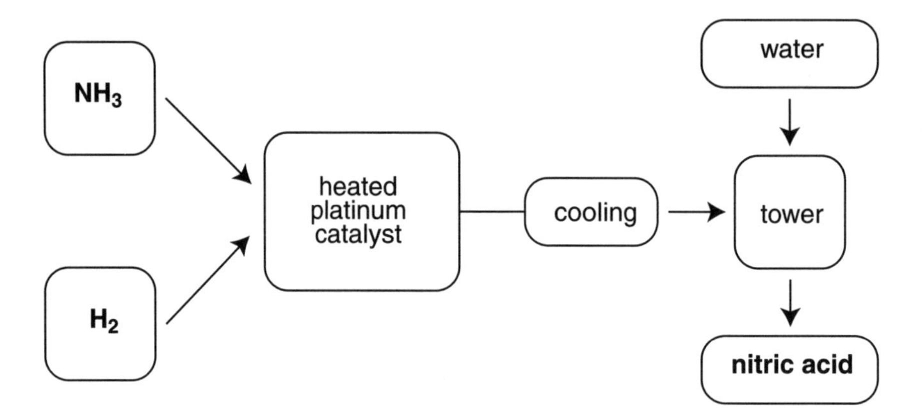

(a) **(i)** How is the ammonia obtained for this process?

.... **[1]**

(ii) Explain the purpose of the platinum catalyst.

.... **[1]**

(iii) The gases leaving the catalyst are cooled before being passed into a tower to react with water.
Suggest how the gases may be cooled.

....

.... **[2]**

(iv) In the water tower nitric acid is produced from nitrogen(IV) oxide, oxygen and water.
Write the equation for this reaction.

.... **[2]**

(b) Nitric acid is reacted with ammonia to make the fertiliser ammonium nitrate.
Ammonium nitrate supplies nitrogen to plants.

(i) Why do plants need nitrogen?

....

.... **[2]**

(ii) What is the percentage by mass of nitrogen in ammonium nitrate?

Leave blank

percentage of ammonia = ... % **[3]**

(iii) Excessive use of ammonium nitrate fertiliser can cause a pollution problem called eutrophication.
Describe and explain this pollution problem.

You will be given credit for a logical sequence in your answer.

....

....

....

....

.... **[4+1]**

(Total 16 marks)

[turn over

Leave blank

8 The table gives information about the first three elements of Group 1 of the Periodic Table.

element	symbol	atomic number	electron arrangement
lithium	Li	3	2,1
sodium	Na	11	
potassium	K		2,8,8,1

(a) These three metals all react vigorously with chlorine to form white crystalline compounds with high melting points.

(i) Draw a dot and cross diagram to show the bonding in the compound between sodium and chlorine.
Only the outer electron shells need to be drawn.

[3]

(ii) Explain why the three metals react with chlorine in such a similar way to produce such similar products.

....

.... **[2]**

(b) The reaction of all three metals with chlorine is vigorous.
These metals will also react with fluorine.

(i) Describe how you would expect these metals to react with fluorine.

.... **[1]**

(ii) Explain your answer to (i).

.... **[1]**

(Total 7 marks)

Leave blank

9 The table shows some bond energies.

bond	bond energy in kJ per mole
C–H	413
O=O	497
C=O	740
O–H	463

Methane reacts with oxygen as shown in the equation.

$$CH_4 + 2O_2 \rightarrow CO_2 + 2H_2O$$

(a) Use the information in the table to calculate the energy change which takes place when one mole of methane is completely burned in oxygen.

[3]

(b) Label the energy level diagram for the reaction.

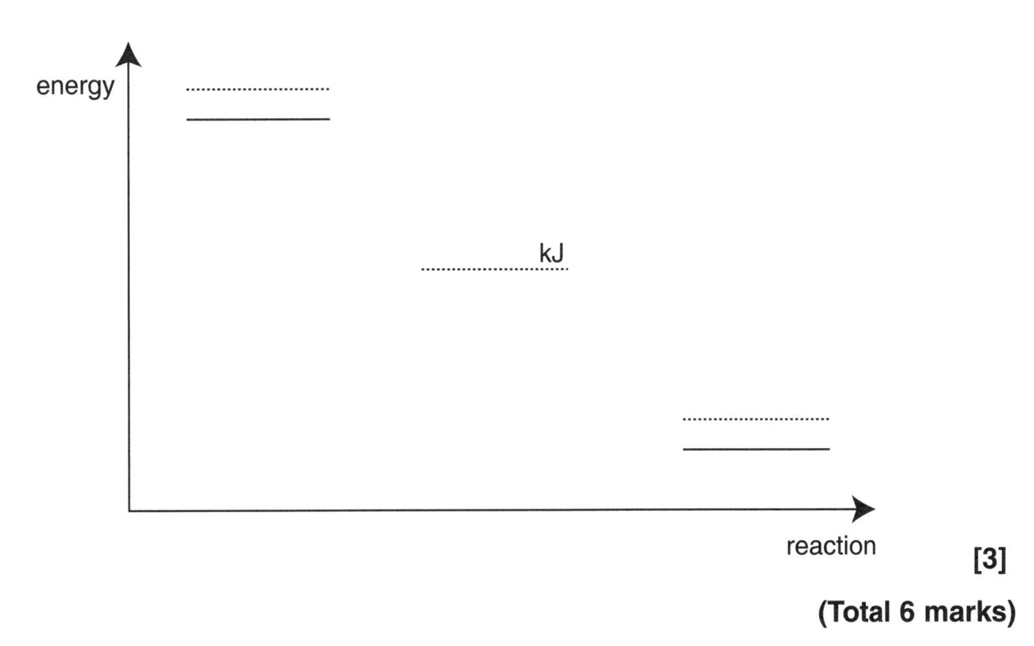

[3]

(Total 6 marks)

BLANK PAGE

Centre number
Candidate number
Surname and initials

Examining Group

General Certificate of Secondary Education

Chemistry
Higher Tier
Exam 1 Paper 2

Time: one hour

Instructions to candidates

Write your name, centre number and candidate number in the boxes at the top of this page.

Answer ALL questions in the spaces provided on the question paper.

Show all stages in any calculations and state the units.
You may use a calculator.

Include diagrams in your answers where this may be helpful.

Information for candidates

The number of marks available is given in brackets **[2]** at the end of each question or part question.

The marks allocated and the spaces provided for your answers are a good indication of the length of answer required.

Where you see this icon you will be awarded marks for the quality of written communication in your answers.
This means, for example, that you should:

- write in sentences
- use correct spelling, punctuation and grammar
- use correct scientific terms.

For Examiner's use only	
1	
2	
3	
4	
5	
Total	

Leave blank

1 Metal oxides react with acids as shown in this general equation.

metal oxide + acid → salt + water

(a) The salt copper(II) chloride can be prepared using this type of reaction.

(i) Write an equation for the reaction which produces this salt.

.... **[2]**

(ii) Describe how you would use this reaction to produce a pure, dry sample of copper(II) chloride.

....

....

....

....

....

....

.... **[4+1]**

(b) Silver chloride is virtually insoluble in water.

(i) What reaction would you use to produce a sample of silver chloride?

....

.... **[1]**

(ii) How would you separate the salt from the reaction mixture?

.... **[1]**

(c) Salts can also be prepared by the reaction between an acid and an alkali.

What name is given to this type of reaction?

.... **[1]**

(Total 10 marks)

Leave blank

2 When chlorine gas is bubbled through a solution of iron(II) sulphate in water, the solution changes from green to brown.
A redox reaction has taken place.

(a) **(i)** What has happened to the chlorine in this reaction?

.... **[1]**

(ii) Describe, in terms of electrons, what has happened to the iron(II) ions in this reaction.

....

.... **[2]**

(iii) Write an ionic equation for the reaction.

.... **[1]**

(b) Describe how you could test solutions containing iron(II) and iron(III) ions to distinguish between them.

....

....

....

.... **[3]**

(c) Mild steel is an alloy of iron.

(i) What is meant by the term alloy?

....

.... **[1]**

(ii) How does the composition of mild steel differ from that of iron?

....

.... **[1]**

[turn over

Leave blank

(iii) Steel is much stronger than iron.
Use ideas of the structure of alloys to explain this.
You may use a diagram to help your answer.

....

....

....

.... **[4]**

(Total 13 marks)

3 **(a)** For industrial use ethanol is made by the reaction between ethene and steam.

$$C_2H_4 + H_2O \rightarrow C_2H_5OH$$

The reaction is carried out at 300°C in the presence of phosphoric acid. Suggest why each of these conditions is used in the process.

300°C ....

phosphoric acid .... **[2]**

(b) In the manufacture of alcoholic drinks, sugar is fermented using yeast to produce ethanol.

(i) Complete this equation for the reaction.

$$C_6H_{12}O_6 \rightarrow .... C_2H_5OH + ...$$ **[2]**

(ii) The reaction is carried out at a temperature of 30°C. Why is this reaction not carried out at 300°C?

....

.... **[2]**

(iii) Why is fermentation not used for the industrial production of ethanol?

....

.... **[2]**

(c) Ethanol can be used as a fuel for motor vehicles.

$$C_2H_5OH + 3O_2 \rightarrow 2CO_2 + 3H_2O$$

(i) Suggest an advantage in the use of ethanol as a fuel for motor vehicles instead of petrol.

.... **[1]**

(ii) What volume of carbon dioxide, measured at room temperature and pressure, would be produced by the complete combustion of 1 kg of ethanol?
(Volume of 1 mol of any gas at room temperature and pressure = 24 dm^3.)

volume = ... dm^3 **[3]**

(Total 12 marks)

Leave blank

[turn over

Leave blank

4 Copper metal can be purified by electrolysis using the apparatus shown in the diagram.

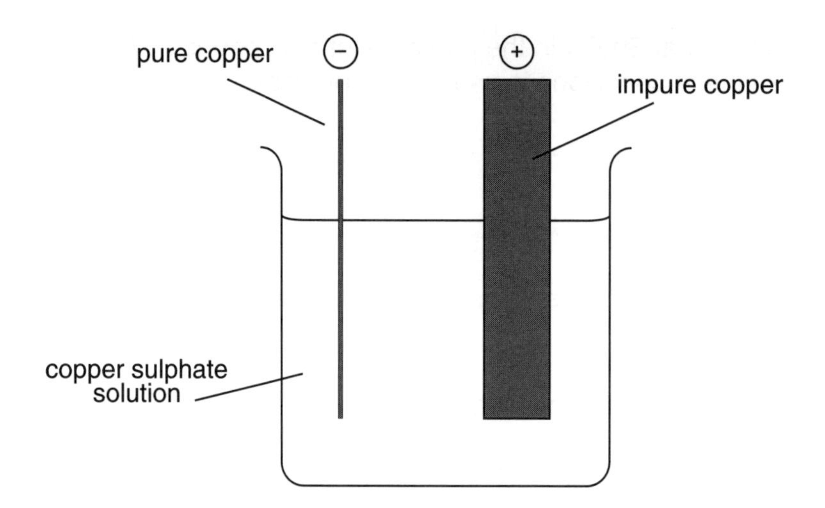

During the electrolysis the positive electrode gets smaller and pure copper is deposited onto the negative electrode.

(a) Suggest a use for very pure copper, and give a reason why high purity is needed for this use.

use ....

reason ....

.... **[2]**

(b) Write an ionic equation for the reaction taking place at each electrode.

(i) At the positive electrode.

.... **[1]**

(ii) At the negative electrode.

.... **[1]**

Leave blank

(c) In this electrolysis a current of 10 A was passed for 20 minutes.
Calculate the mass of pure copper deposited onto the negative electrode.
(Faraday constant = 96 500 coulombs.)

mass = ... g **[3]**

(d) Two of the impurities in the positive electrode are zinc and platinum.
Suggest what happens to these two impurities during the electrolysis.

....

....

....

....

....

.... **[4]**

(e) Copper metal is a good conductor of electricity.
Use your knowledge of the structure of metals to explain how copper conducts electricity.

You will be given credit for spelling, punctuation and grammar.

....

....

....

....

.... **[3+1]**

(Total 15 marks)

5 The time taken for each of four different gaseous elements to diffuse the length of a glass tube was measured. These measurements are given in the table, together with more information about the elements.

gaseous element	atomic number	average atomic mass	diffusion time in seconds
argon	18	40	3.10
hydrogen	1	1	0.70
oxygen	8	16	2.72
xenon	54	131	5.71

(a) (i) Describe what happens in the process of diffusion.

....

....

.... **[3]**

(ii) Explain why xenon has the slowest rate of diffusion.

....

.... **[2]**

(b) Some elements have atoms with different atomic masses.
They are called isotopes.

(i) Explain what is meant by the term 'isotope'.

....

.... **[2]**

(ii) The table gives information about the isotopes of the element chlorine.

isotope	relative mass	percentage by mass
^{35}Cl	37	25
^{37}Cl	35	75

Use the information in the table to calculate the average mass of a chlorine atom.

average mass = ... **[3]**

(Total 10 marks)

Leave blank

Letts

Centre number
Candidate number
Surname and initials

Letts Examining Group

General Certificate of Secondary Education

Chemistry
Higher Tier
Exam 2 Paper 1

Time: one and a half hours

Instructions to candidates

Write your name, centre number and candidate number in the boxes at the top of this page.

Answer ALL questions in the spaces provided on the question paper.

Show all stages in any calculations and state the units.
You may use a calculator.

Include diagrams in your answers where this may be helpful.

Information for candidates

The number of marks available is given in brackets **[2]** at the end of each question or part question.

The marks allocated and the spaces provided for your answers are a good indication of the length of answer required.

Where you see this icon you will be awarded marks for the quality of written communication in your answers.
This means, for example, that you should:

- write in sentences
- use correct spelling, punctuation and grammar
- use correct scientific terms.

For Examiner's use only	
1	
2	
3	
4	
5	
6	
7	
8	
9	
10	
Total	

Leave blank

1 Copper(II) oxide is added to dilute sulphuric acid in a test tube and the mixture warmed. When the test tube is left in a rack for a few minutes a black solid settles to the bottom, leaving a clear blue liquid.

(a) **(i)** Write down the name of the clear blue liquid.

.... **[1]**

(ii) Write down the name of the black solid in the test tube.

.... **[1]**

(iii) Write a balanced equation for the reaction between copper(II) oxide and sulphuric acid.

.... **[2]**

(iv) What name is given to this type of reaction?

.... **[1]**

(b) Describe how you would make blue crystals from the mixture in the test tube. (One mark is for the correct sequencing of your answer.)

....

....

....

.... **[3+1]**

(c) The reaction between copper(II) oxide and sulphuric acid gives out energy in the form of heat.

(i) What name is given to a reaction which gives out heat?

.... **[1]**

(ii) Draw an energy level diagram to represent this reaction.

[3]

(Total 13 marks)

2 The apparatus shown in the diagram was used to study the combustion of a liquid hydrocarbon, octane, C_8H_{18}.

Leave blank

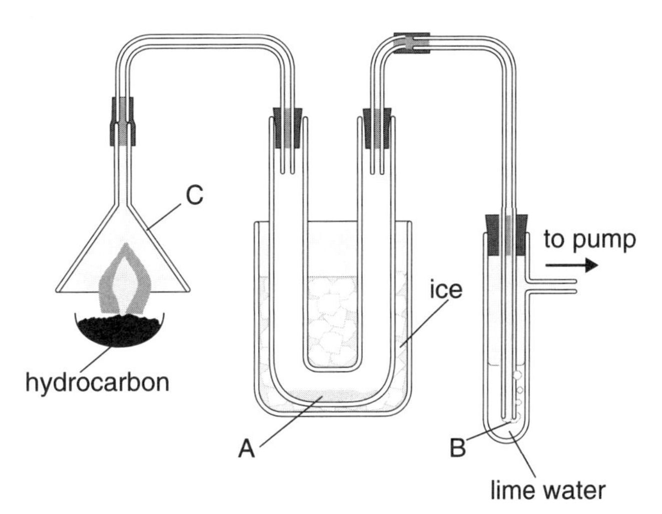

Gases from the burning hydrocarbon were drawn through the apparatus for several minutes.

(a) A clear, colourless liquid appeared at A.

(i) Name this liquid.

.... **[1]**

(ii) Describe a test to prove the identity of this liquid.

....

.... **[2]**

(b) (i) What would you see at B as the experiment was carried out?

....

.... **[2]**

(ii) What does this show about the gases produced by combustion of the hydrocarbon?

.... **[1]**

(c) (i) Write a balanced equation for the complete combustion of octane.

.... **[2]**

(ii) When the experiment was completed, a black deposit of carbon was noted at C.
Explain how this was formed.

....

.... **[2]**

(Total 10 marks)

[turn over

3 Sarah studied the reaction between hydrochloric acid and sodium carbonate.

Leave blank

$$2HCl + Na_2CO_3 \rightarrow 2NaCl + CO_2 + H_2O$$

She made hydrochloric acid of different concentrations by mixing a more concentrated solution with water. She used tablets each of which contained the same mass of sodium carbonate. She timed how long it took for a tablet to react completely in the same volume of each concentration of acid.
Her results are shown in the table.

volume of acid in cm^3	volume of water in cm^3	time for tablet to react in seconds
2	18	350
4	16	245
6	14	220
8	12	142
10	10	57

(a) Plot the volume of acid used against time on the grid below.
Draw the line of best fit for the points you have plotted. **[3]**

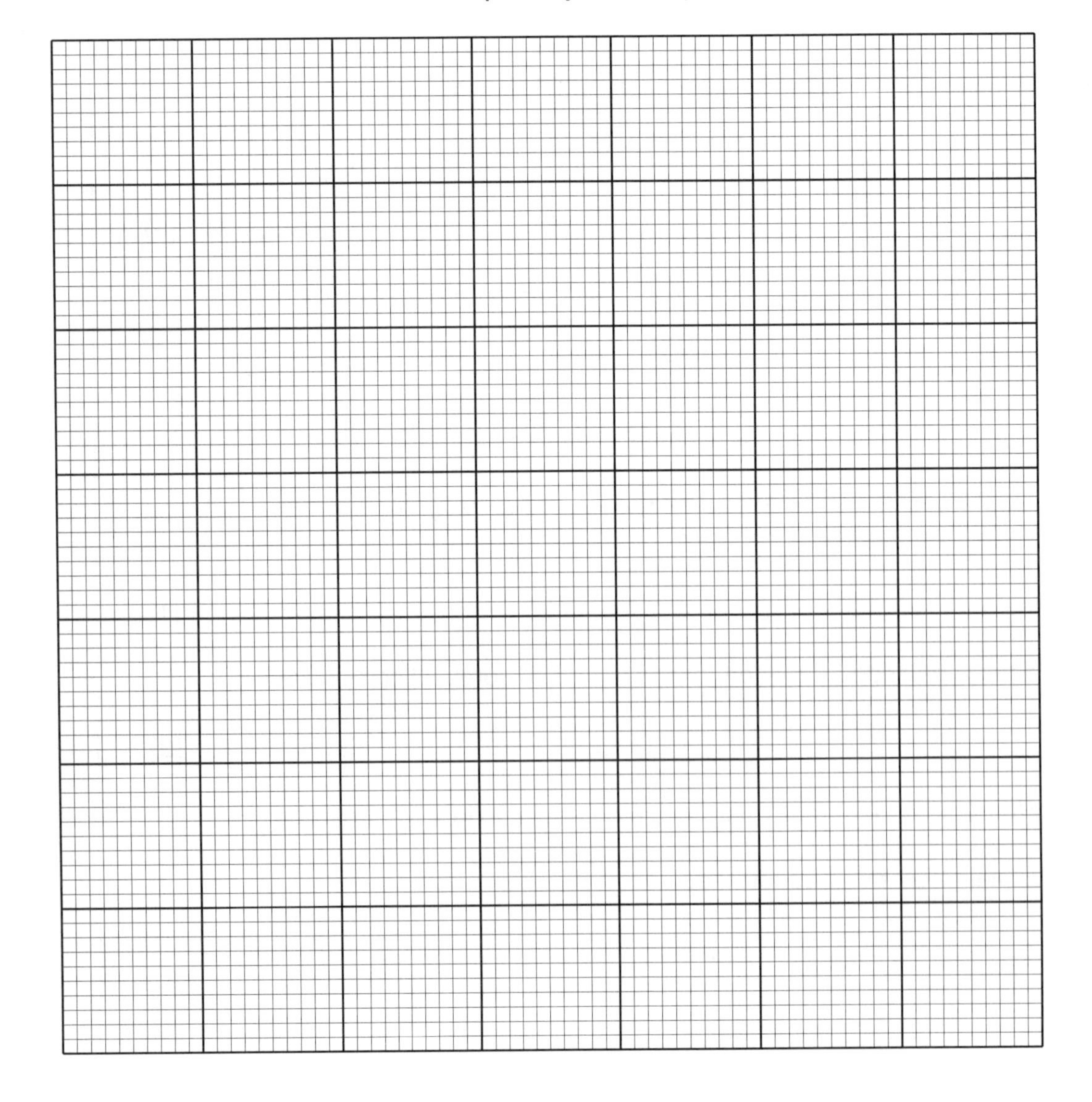

(b) Sarah was careful to ensure fair testing in her experiments.

Leave blank

(i) Explain how mixing each volume of acid with a different volume of water helped to ensure fair testing.

....

....

.... [2]

(ii) What other thing, not mentioned above, must Sarah have kept constant to ensure fair testing?

.... [1]

(c) One of Sarah's results is anomalous.

(i) What volume of acid was used for the anomalous result?

.... [1]

(ii) Suggest what may have caused the error in this result.

....

.... [1]

(d) (i) Describe the relationship between concentration of acid and rate of this reaction shown by Sarah's results.

....

....

.... [2]

(ii) Use your knowledge of particles to explain this relationship.

....

....

....

.... [3]

(Total 13 marks)

[turn over

Leave blank

4 The apparatus shown below was used to find the formula of an oxide of copper.

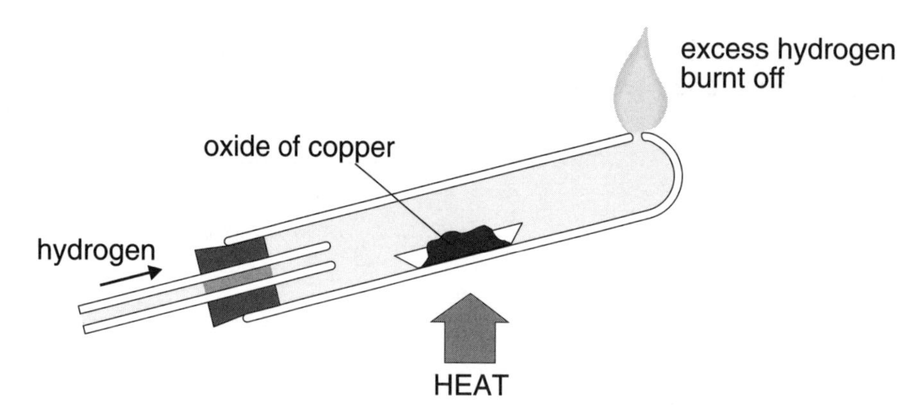

Hydrogen gas was passed over a heated sample of the oxide of copper in a ceramic container. The oxide of copper was reduced to copper metal. The following masses were measured.

mass of ceramic container = 12.64 g
mass of ceramic container + oxide of copper = 15.88 g
mass of ceramic container + copper = 15.52 g

(a) (i) What mass of the copper was formed in the experiment?

.... **[1]**

(ii) What mass of oxygen was combined with this mass of copper?

.... **[1]**

(b) Use the values from (a) to work out the formula of this oxide of copper.
(Relative atomic masses: Cu = 64, O = 16)

[4]

(c) This method could not be used to find the formula of sodium oxide.
Explain why.

....

.... **[2]**

(Total 8 marks)

Leave blank

5 The apparatus below was used to break the large hydrocarbon molecules in petroleum jelly into smaller hydrocarbon molecules.

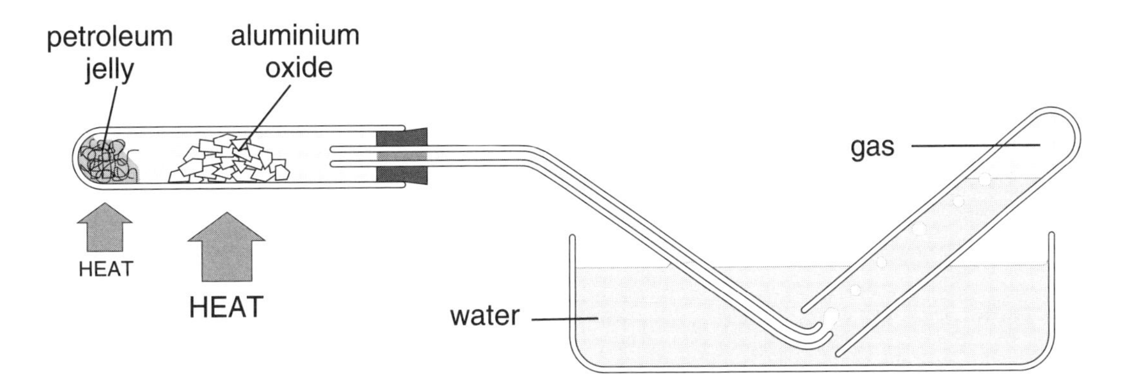

(a) **(i)** What name is given to this type of reaction?

....

.... **[2]**

(ii) What is the purpose of the aluminium oxide?

....

.... **[1]**

(iii) The aluminium oxide is in small pieces.
Explain how the use of small pieces of aluminium oxide speeds up the reaction more than the use of large pieces would.

....

....

....

.... **[2]**

[turn over

Leave blank

(b) The gas collected in the tube has a molecule containing two carbon atoms.

(i) Name the gas collected in the tube.

.... **[1]**

(ii) Describe how you would test this gas to show that it is not an alkane.

....

....

.... **[2]**

(c) The alkane decane, $C_{10}H_{22}$, can be broken down in a similar reaction to give octane, C_8H_{18}, and the gas in (b).
Draw graphical (displayed) formulae to show the equation for this reaction.

[3]

(Total 11 marks)

6 This diagram shows a cross section through some layers of rock in the Earth's crust.

Leave blank

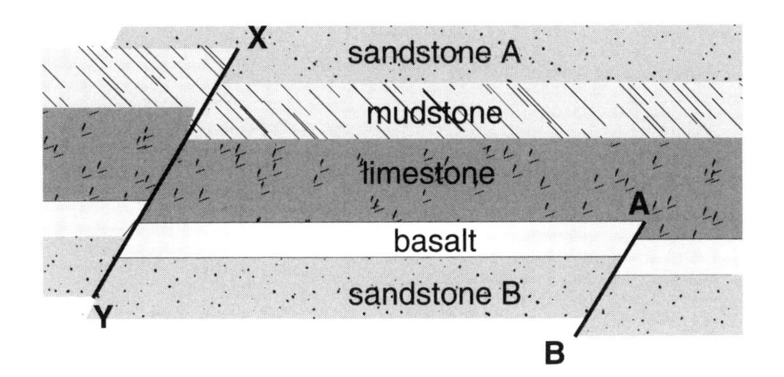

(a) Which of the rocks shown in the diagram:

(i) is the youngest sedimentary rock? .... **[1]**

(ii) is an igneous rock? .... **[1]**

(iii) is mostly calcium carbonate? .... **[1]**

(b) The basalt is made up from small crystals.
Explain what this tells you about how it was formed.

....

....

.... **[2]**

(c) How can you tell that the fault A–B occurred before the fault X–Y?

....

....

.... **[2]**

(Total 7 marks)

Letts

7 The diagram shows the carbon cycle.

Leave blank

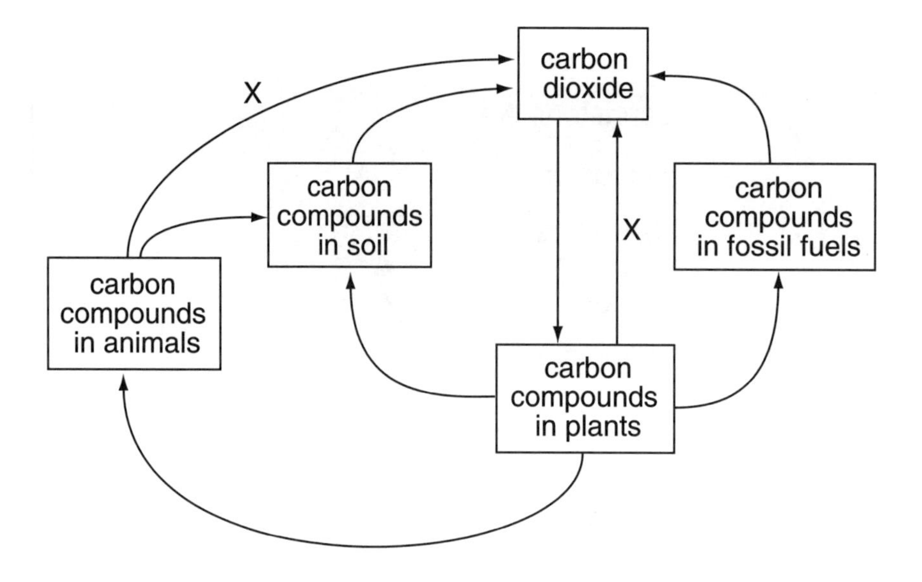

(a) **(i)** Write down the name of the process marked X.

.... **[1]**

(ii) Write a balanced equation for the overall process of photosynthesis.

.... **[2]**

(iii) The process of photosynthesis requires energy.
Explain how this energy is obtained.

....

.... **[2]**

(b) The percentage of carbon dioxide in the Earth's atmosphere remained constant for thousands of years until the twentieth century.

(i) Use the carbon cycle to explain how the percentage of carbon dioxide in the Earth's atmosphere remained constant.

....

....

....

.... **[3]**

Leave blank

(ii) Describe and explain the change in percentage of carbon dioxide in the Earth's atmosphere which occurred during the twentieth century.

....

....

.... **[2]**

(c) Four billion years ago the Earth's atmosphere contained hydrogen, helium and ammonia.
Suggest how these gases were removed from the atmosphere.
(One mark for correct spelling, punctuation and grammar.)

....

....

....

....

.... **[4+1]**

(Total 15 marks)

8 Ammonia is manufactured from nitrogen and hydrogen.

Leave blank

$$N_2(g) + 3H_2(g) \rightleftharpoons 2NH_3(g)$$

(a) What does the symbol $\rightleftharpoons$ show?

.... **[1]**

(b) Where are the raw materials nitrogen and hydrogen obtained from?

nitrogen is obtained from ....

hydrogen is obtained from .... **[2]**

(c) Suggest two reasons why the manufacture of ammonia is carried out at high pressure.

1 ....

2 .... **[2]**

(d) At low temperatures a very high yield of ammonia can be obtained if the mixture is left long enough. Explain why the process is actually carried out at higher temperatures which give a lower yield.
Use ideas about collisions between particles in your answer.

....

....

.... **[3]**

(e) Why was the discovery of a method to make ammonia on a large scale globally important?

....

....

.... **[2]**

(f) Calculate the mass of ammonia that could be produced if 28 tonnes of nitrogen is completely converted into ammonia.
(Relative atomic masses: H = 1, N = 14.)

mass = ... tonnes **[2]**

(Total 12 marks)

9 The diagram below shows part of the Periodic Table.

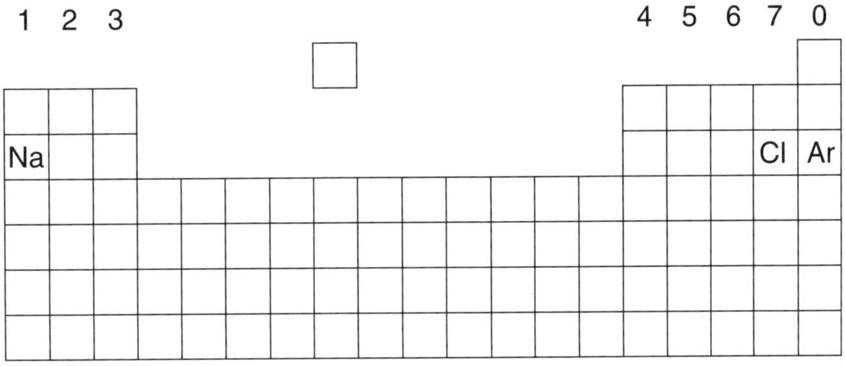

Leave blank

1 2 3 4 5 6 7 0

Na Cl Ar

The position of three elements in the Periodic Table is shown.

(a) Describe the difference in the atomic structure of these three elements.

....

....

....

.... **[2]**

(b) Using these three elements as examples, describe the trend in chemical properties across the second period of the Periodic Table.

....

....

....

.... **[3]**

(c) Use ideas about the electronic structure of the three elements to explain this trend in chemical properties.

....

....

.... **[2]**

(Total 7 marks)

Leave blank

10 The table shows the composition of the ocean.

Ion	Concentration of ion in g/100g of sea water
Chloride Cl^-	19.2
Sodium Na^+	10.7
Sulphate SO_4^{2-}	2.7
Magnesium Mg^{2+}	1.4
Calcium Ca^{2+}	0.4
Potassium K^+	0.38
Hydrogencarbonate HCO_3^-	0.14
Bromide Br^-	0.07

Rocks such as halite (sodium chloride) and calcite (calcium sulphate) are dissolved in rivers and enter the oceans. Other rocks such as limestone, chalk or marble react with rain water and get washed into rivers.

(a) Suggest why there are low concentrations of calcium ions in the ocean despite large quantities of calcium compounds in river water.

....

....

.... **[2]**

(b) Write an ionic equation to support your answer to (a)

.... **[2]**

(Total 4 marks)

Letts

BLANK PAGE

BLANK PAGE

Centre number
Candidate number
Surname and initials

Letts Examining Group

General Certificate of Secondary Education

Chemistry
Higher Tier
Exam 2 Paper 2

Time: one hour

Instructions to candidates

Write your name, centre number and candidate number in the boxes at the top of this page.

Answer ALL questions in the spaces provided on the question paper.

Show all stages in any calculations and state the units.
You may use a calculator.

Include diagrams in your answers where this may be helpful.

Information for candidates

The number of marks available is given in brackets **[2]** at the end of each question or part question.

The marks allocated and the spaces provided for your answers are a good indication of the length of answer required.

Where you see this icon you will be awarded marks for the quality of written communication in your answers.
This means, for example, that you should:

- write in sentences
- use correct spelling, punctuation and grammar
- use correct scientific terms.

For Examiner's use only	
1	
2	
3	
4	
5	
6	
Total	

1 The table gives some information about the homologous series of alkanes.

Leave blank

name	formula	molecular mass	boiling point in°C
methane	CH_4	16	–161
ethane	C_2H_6	30	....
....	C_3H_8	44	–42
butane	....	58	–1
pentane	C_5H_{12}	72	36

(a) Complete the table by filling in the three blank boxes. **[3]**

(b) Explain what is meant by the term homologous series as it applies to the alkanes.

....

....

.... **[2]**

(c) Butane exists as a number of structural isomers.

(i) What are structural isomers?

....

.... **[2]**

(ii) Draw structural (displayed) formulae for **two** structural isomers of butane.

[2]

(Total 9 marks)

Leave blank

2 Simon investigated the solubility of potassium nitrate in water.
He measured the mass of potassium nitrate which would dissolve in $100\,cm^3$ of water at different temperatures.
His results are shown in the table.

temperature in °C	solubility of potassium nitrate in g per $100cm^3$ water
20	32
30	47
40	63
50	65
60	110
70	138

(a) Plot the solubility of potassium nitrate against temperature on the grid below. Draw the line of best fit for the points you have plotted. **[3]**

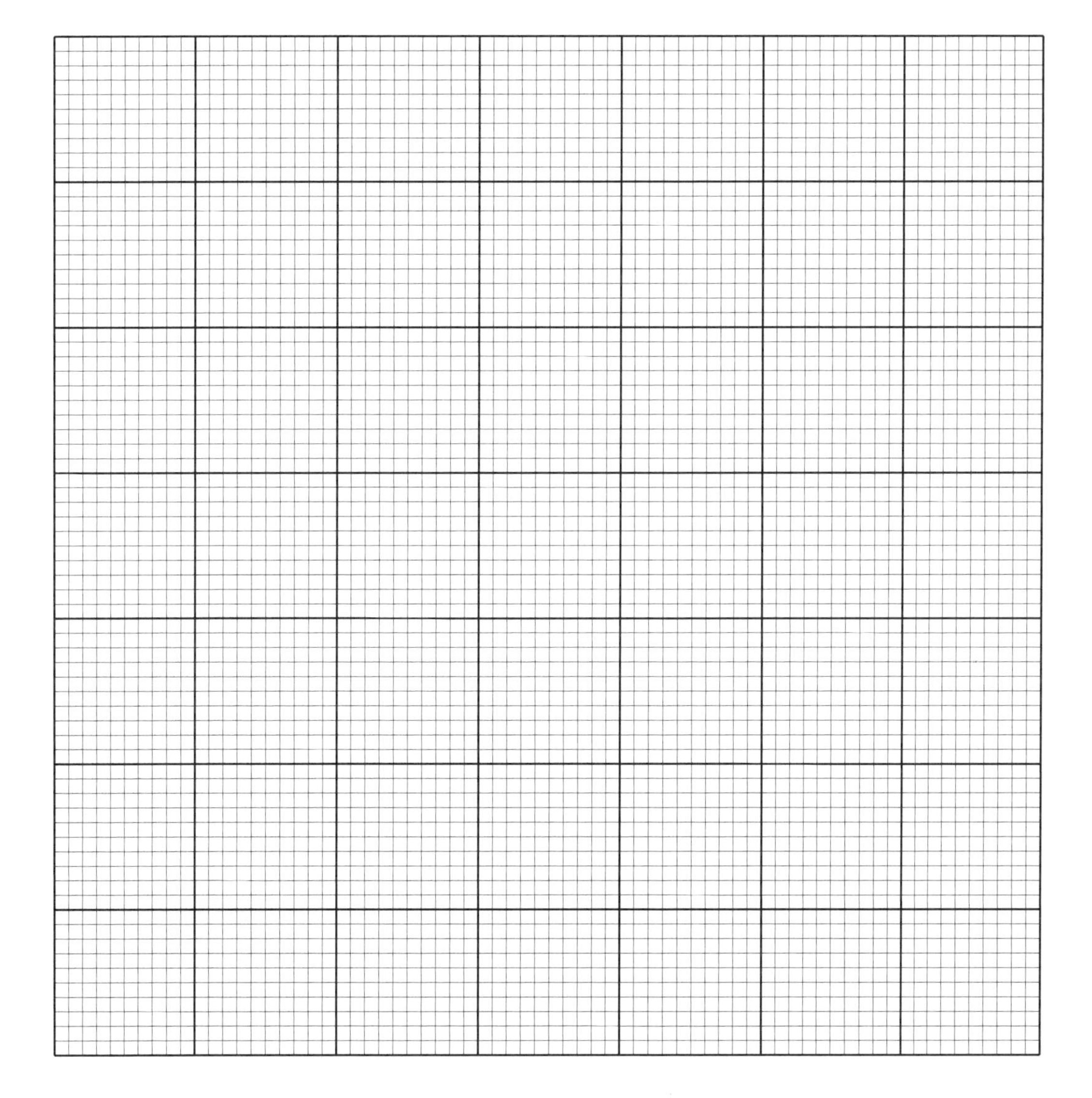

[turn over

(b) Describe the way in which the solubility of potassium nitrate changes with temperature.

Leave blank

....

....

.... [2]

(c) One of Simon's results is anomalous.

(i) At which temperature did the anomalous result occur?

.... [1]

(ii) Suggest the correct value for the solubility at this temperature.

.... [1]

(d) (i) Describe what you would see if the solution Simon made at 70°C was cooled slowly to room temperature.

....

.... [2]

(ii) Explain your answer to (i).

....

.... [2]

(Total 11 marks)

3 Some coins are made of an alloy of zinc, nickel and copper. To find the percentage of zinc in the coins one coin, of mass 0.5 g, was placed in 25 cm^3 of hydrochloric acid of concentration 0.5 mol/dm^3. Only the zinc reacted.

$$Zn + 2HCl \rightarrow ZnCl_2 + H_2$$

When the reaction had finished the mixture was filtered and titrated against sodium hydroxide solution of 0.5 mol/dm^3 concentration.
To reach neutralisation point took 14.6 cm^3 of this sodium hydroxide solution.

(a) (i) How could you see when the reaction between the zinc and hydrochloric acid had finished?

.... **[1]**

(ii) Explain why the zinc reacted with the hydrochloric acid, but the nickel and copper did not.

....

.... **[1]**

(b) (i) Calculate the volume of 0.5 mol/dm^3 hydrochloric acid which reacted with 14.6 cm^3 of 0.5 mol/dm^3 sodium hydroxide solution.

volume = ... cm^3 **[2]**

(ii) Calculate the volume of 0.5 mol/dm^3 hydrochloric acid which reacted with the zinc.

volume = ... cm^3 **[1]**

Leave blank

[turn over

Leave blank

(iii) Calculate the mass of zinc which reacts with this volume of 0.5 mol/dm^3 hydrochloric acid.
(Relative atomic mass: Zn = 65)

mass = ... g **[3]**

(iv) What percentage of zinc was in the coins?

percentage = ... % **[2]**

(Total 10 marks)

Leave blank

4 The apparatus shown in the diagram was used to electroplate an iron medallion of mass 10.50 g with a metal M.
The metal M has a relative atomic mass of 59.

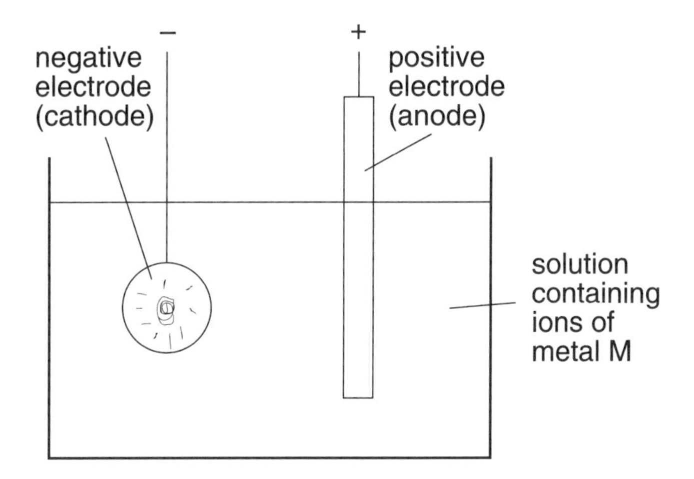

A current of 2 amperes was passed for 12 minutes.
After electroplating the medallion had a mass of 10.94 g.

(a) (i) Suggest one advantage there could be in electroplating the iron medallion.

....

.... **[1]**

(ii) Suggest a suitable material for the positive electrode (anode) in the apparatus.

.... **[1]**

(iii) Explain your choice in (ii).

....

....

.... **[1]**

(b) Calculate the following quantities. (1 Faraday = 96 500 coulombs.)

(i) The mass of M electroplated onto the medallion.

mass of M = ... g **[1]**

[turn over

Leave blank

(ii) The quantity of electricity, in coulombs, passed through the circuit during the electroplating.

quantity of electricity = ... C **[2]**

(iii) The number of positive charges on an ion of the metal M.

number of positive charges = ... **[2]**

(Total 8 marks)

Leave blank

5 The table gives information about the atomic structure of some elements.

element	protons	arrangement of electrons	Group in the Periodic Table	metal or non-metal
Q	8	2, 6	6	....
X	....	2, 8, 1	1	metal
Y	17	....	7	non-metal
Z	18	2, 8, 8	....	non-metal

(a) Complete the table by filling in the four blank boxes. **[4]**

(b) Element X will react with another of the elements in the table to form a crystalline salt.
Write down the letter of this other element.

.... **[1]**

(c) Element X forms a compound with element Q.
Use the letters X and Q to show the formula of this compound.

.... **[1]**

(d) Element X will conduct electricity.
Use your knowledge of the structure of metals to explain this property.
Use a diagram to help your answer.

....

....

.... **[3]**

(e) Element Y is diatomic. Explain what this means.

....

.... **[2]**

(Total 11 marks)

Letts

6 Water hardness can be shown as the concentration of calcium ions in solution. The table shows the calcium ion concentration of samples of water from three different sources before and after boiling or addition of sodium carbonate.

Leave blank

		calcium ion concentration parts per million	
source	**treatment**	**before treatment**	**after treatment**
A	boiled	27	15
A	sodium carbonate	27	0
B	boiled	23	0
B	sodium carbonate	23	0

(a) (i) Explain how boiling removes hardness from water.
(One mark is for the correct use of scientific language.)

....

....

....

.... **[3+1]**

(ii) Boiling removed all of the hardness from the water from source B, but only part of the hardness from the water from source A.
Suggest a reason for this.

....

.... **[1]**

(b) The addition of sodium carbonate removed all of the hardness from the water from both sources.

(i) Explain how sodium carbonate removes hardness from water.
(One mark is for correct spelling, punctuation and grammar.)

....

....

.... **[3+1]**

(ii) Explain why using sodium carbonate removed all of the hardness from the water from both sources.

....

.... [1]

(c) Suggest a problem which might be caused by water hardness in the homes of people using water from source A.

....

....

.... [1]

(Total 11 marks)

Leave blank

Letts Educational
The Chiswick Centre
414 Chiswick High Road
London
W4 5TF

Tel: 020 8996 3333
Fax: 020 8742 8390
Email: mail@lettsed.co.uk

First published 2004

Prepared by *specialist* publishing services, Milton Keynes

British Library Cataloguing in Publication Data

A CIP record for this title is available from the British Library

ISBN 1843153068

Letts Educational Limited is a division of Granada Learning Limited, part of Granada plc.

Printed in the UK

Letts Examining Group
General Certificate of Secondary Education
Chemistry Higher Tier
Mark scheme and Examiner's report

Answers: GCSE Chemistry exam 1 paper 1

Question	Answer	Mark
1 a	A	1
b	fuel oil	1

Examiner's Tip
At the base of the fractionating tower very hot oil vapour enters. The vapour cools as it passes up the tower, so the coolest part is at the top. Fractions condense from the vapour as it cools, so the highest boiling point fraction, with the largest molecules, condenses first – at the bottom.

c i

```
      H   H
      |   |
  H — C — C — H
      |   |
      H   H
```

2

Question	Answer	Mark
c ii	alkanes	1

Examiner's Tip
There are two families or homologous series of hydrocarbons that you need to know about. The first contains only single covalent bonds and is called the alkanes. Ethane belongs to this group. The second group contains a double covalent bond between two carbon atoms, and is called the alkenes.

Question	Answer	Mark
d i	$C_2H_6 \rightarrow H_2 + C_2H_4$	1
ii	ethene	1
iii	Shake the alkene with bromine water.	1
	The orange/bromine water turns colourless.	1

Examiner's Tip
The second product of this reaction is an alkene. Just like the alkanes, the name of each alkene is based on the number of carbon atoms it contains. Since this one contains two carbon atoms its name is ethene. An alkene with three carbon atoms would be propene, etc.
Alkenes will decolourise bromine water. As the alkene reacts, one of the two bonds between the carbon atoms breaks and the bromine atoms join on. An alkane will not react with bromine in this way.

e i

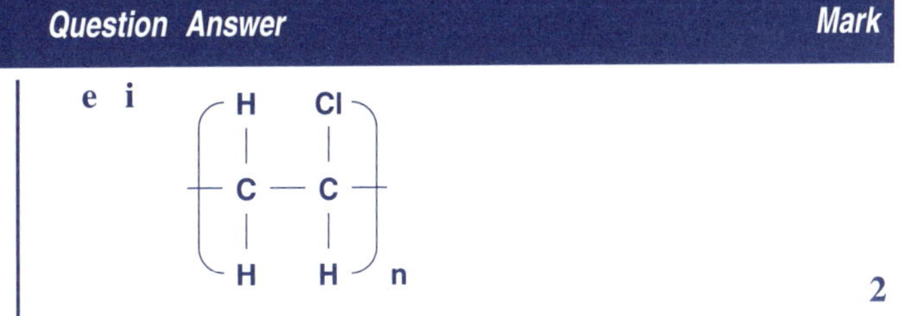

2

Examiner's Tip
The easiest way to show the formula of a polymer is to draw the repeating unit, which comes from one molecule of the monomer, place brackets around it, and indicate that this repeats by writing the letter n outside the right bracket. The letter n stands for a large number of this unit repeated down the length of the polymer.

Question	Answer	Mark
ii	use: e.g. substitute for leather	1
	property: e.g. flexible	1
2 a i	Hydrogen and helium are gases with very low density	1
	and diffused off into space.	1

Examiner's Tip
Only the gases with higher densities have been held in the Earth's atmosphere by gravity. It would be acceptable to say that hydrogen and helium have very small or very light atoms and could not be held in the atmosphere by gravity.

Question	Answer	Mark
ii	The atmosphere did not contain oxygen.	1
b	The first plants appeared and carried out photosynthesis.	1
	This released oxygen into the atmosphere	1
	and used carbon dioxide from the atmosphere.	1
	Flammable gases like methane burned in the oxygen.	1
	Bacteria converted ammonia to nitrogen.	1
	+ 1 mark for spelling, punctuation and grammar	1

Examiner's Tip
Most of the changes to the Earth's atmosphere occurred because of photosynthesis. Over a long period of time this reduced the carbon dioxide content and increased the oxygen content.

Question	Answer	Mark
3 a i	Li and Be are solid but N and O are gases.	1
ii	Li and Be have strong metallic bonding.	1
	They contain ions with a sea of electrons between them.	1
	N and O are made of molecules with covalent bonds between their atoms.	1
	These molecules have weak bonding between them.	1

Examiner's Tip
Use information from the table to help your answer. Look carefully at the boiling points. It is easy to become confused when some of the figures are negative ones.
Metals like lithium and beryllium contain positive ions which are held together by strong attraction to the negative electrons, which form a 'sea' between the ions. Non-metals like nitrogen and oxygen have pairs of atoms joined together to make molecules. The bonds within these molecules are very strong covalent bonds, but between the molecules are weak van der Waals forces. The strong bonds in metals need a lot of heat energy to be broken, but weak forces between molecules are broken at low temperatures.

Question	Answer	Mark
b	Diamond has a giant covalent structure.	1
	A very large quantity of energy is needed to break all of the covalent bonds joining the atoms.	1
c i	Si	1
ii	Si is in the same group as carbon.	1
	Elements in the same group usually have similar physical properties.	1

Examiner's Tip
With a few exceptions, elements in the same group of the Periodic Table have similar physical properties. These properties are more similar when the elements are next to each other in the group. In Group 4 carbon is the first member and silicon the second. They both have giant covalent structures and therefore similar high boiling points.

Question	Answer	Mark
d i	Boiling points increase down the group.	1
ii	Down the group the atoms are heavier;	1
	heavier atoms need more energy to boil.	1
4 a	The reaction is exothermic.	1
	The reaction produces enough heat energy to melt the iron.	1
b	The burning magnesium provides enough energy to start the reaction.	1
	this is known as the activation energy for the reaction	1
c i	The iron in the oxide will only be displaced by a more reactive metal.	1
	Copper is below iron in the reactivity series.	1
ii	magnesium	1
iii	a reaction involving both reduction and oxidation	1

Examiner's Tip
In displacement reactions, a more reactive metal takes the place of a less reactive metal. You need to know where the common metals are in the reactivity series to predict what will happen in displacement reactions.
In this reaction the iron in iron oxide is reduced to iron metal and the aluminium metal is oxidised to aluminium oxide. This means it is a redox reaction.

Question	Answer	Mark
d	greater surface area	1
	gives higher rate of reaction	1
e	$2Al + Fe_2O_3 \rightarrow Al_2O_3 + 2Fe$ $(2\times56)+(3\times16)=160g\ Fe_2O_3$ produces $2\times56=112gFe$ $500g\ Fe_2O_3$ produces $500\times{}^{112}/_{100}g\ Fe\ =350g$	3

Examiner's Tip
You must use the equation to work out the reacting quantities of aluminium and iron(III) oxide, and the iron produced. To do this, use the number of each formula in the equation and the masses of the formulae concerned, calculated by adding atomic masses.
From the quantities involved it is clear that the 500 g of iron(III) oxide would run out before all of the aluminium was used up. You must therefore use the iron(III) oxide to calculate how much iron is produced.

5 a

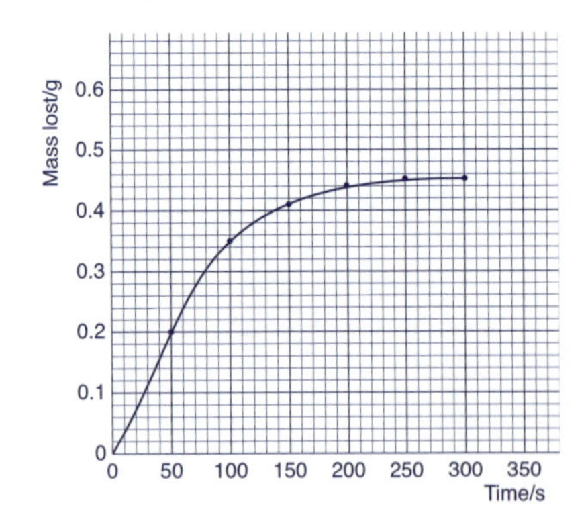

Examiner's Tip
Mark each point you plot clearly with a dot or cross. Draw the best fit line ignoring any anomalous points. You must draw one continuous line in a smooth curve. In this case no mass would be lost at zero time, so you must draw your curve starting at 0,0.

Question	Answer	Mark
	all six points plotted correctly to within $\frac{1}{2}$ small square	2
	best fit curve drawn through 0,0 to horizontal line at mass 0.45 ignoring point at 50,0.2	1
b	The reaction produces carbon dioxide which bubbles off from the mixture.	1

Question	Answer	Mark

c

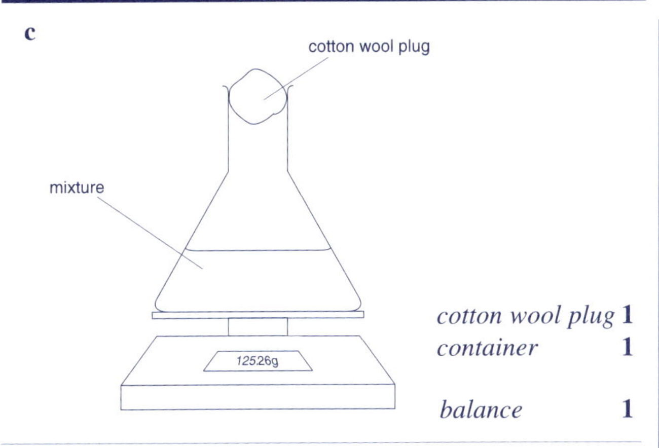

cotton wool plug 1
container 1
balance 1

Examiner's Tip
Since a gas is given off in this reaction, the mass remaining in the mixture will decrease with time. This can be measured using a balance. Any suitable container can be used, but it should have a cotton wool ball in the mouth to prevent liquid splashing out as the mixture bubbles.

d i *second graph with steeper slope* 1
but with same maximum of 0.45 g lost 1

ii powder has greater surface area 1
allowing more contact between solid and liquid 1
so giving a higher rate of reaction 1

Examiner's Tip
Don't forget to label the new graph on the grid. An easy mistake to make is to draw this second curve to a different maximum value. Since the same mass of copper (II) carbonate has been used, the total mass lost will be the same as the first graph.

6 a i The gas was chlorine 1
which bleaches colours. 1

ii $2Cl^- \rightarrow Cl_2 + 2e^-$
one mark for each side of the equation 2

Examiner's Tip
Chloride ions are negative and so are attracted to the positive electrode, where they give up electrons and become chlorine gas. Do not forget that chlorine is diatomic, that is, it exists as molecules containing pairs of atoms. Chlorine is a powerful oxidising agent, which will bleach coloured pigments to colourless, so the litmus loses its blue colour to become white paper.

b i hydrogen 1

ii the sodium chloride solution contains the ions Na^+, Cl^-, H^+ and OH^- 1
hydrogen ions are being lost from the negative electrode 1
this leaves hydroxide, OH, ions 1
which make the solution alkaline 1

Question	Answer	Mark

c *two from*:
chlorine used in swimming pools/tap water
hydrogen used to make ammonia/margarine
sodium hydroxide used in oven cleaners 2

7 a i by the Haber process 1

ii to speed up the reaction 1

iii using water 1
passed on the outside of pipes through which the hot gases pass 1

iv $4NO_2 + O_2 + 2H_2O \rightarrow 4HNO_3$
one mark lost for each error 2

b i to make protein 1
for growth 1

ii NH_4NO_3
mass of nitrogen = 2 × 14 = 28 1
formula mass = 14 + 4 + 14 + 48 = 80 1
% nitrogen = 100 × 28/80 = 35 1

Examiner's Tip
To calculate the percentage of an element in a compound, first write out the formula. Then work out the mass of the element in the formula and the formula mass. The percentage can then be calculated.

iii *four from*:
Rain washes excess fertiliser into streams/rivers/lakes.
Using this fertiliser algae grow rapidly.
The algae block out sunlight.
Plants lower in the water die and rot.
Bacteria feeding on dead plants increase in number. Large numbers of bacteria use up the oxygen in the water. Without oxygen fish die. 4
+ 1 mark for a logical sequence 1

8 a i

```
     o o                 o x
 o         o          x       x
 o   Na+   o          x  Cl-  x
     o o                 x x
```

3

Examiner's Tip
Usually dot and cross diagrams only need to show the outer electrons. Dots, or small circles, are used to show the electrons from the one atom, and crosses the electrons from the other atom. In this case, both of the ions formed have eight electrons in the outer shell, but note that chlorine has one electron which has been gained from the sodium atom as the ions were formed. This means that the outer shell of the sodium atom, which only contained one electron, has gone. The second shell now becomes the outer shell. Don't forget to mark which ion is which, using the correct symbol and charge.

Question	Answer	Mark

ii The atoms of these three metals each have one electron in the outer shell. **1**
This is transferred to a chlorine atom to form an ionic bond. **1**

Examiner's Tip
Elements from the same group of the Periodic Table react in the same way because they have similar electron arrangements. Look at the example given in the question to see how the electrons are forming bonds – in this case ionic bonds are formed by electron transfer. Elements from the same group will transfer the same number of electrons and therefore behave in a similar way to form similar compounds.

b i more vigorously than they react with chlorine **1**

ii In Group VII reactivity decreases down the group, so fluorine is more reactive than chlorine. **1**

Examiner's Tip
You need to learn that for Group VII, the halogens, reactivity of the elements decreases down the group. For Group I, the alkali metals, reactivity increases down the group. Make sure you don't mix up the two groups.

9 a

$CH_4 + 2O_2$	$\rightarrow$	$CO_2 + 2H_2O$	
bonds broken	energy used	bonds made	energy released
4 C–H	4×413=1652	2 C=O	2×740=1480
2 O=O	2 × 497 = 994	4 H–O	4 × 463 = 1852
total energy used = 2646		total energy released = 3332	

difference = 3332 − 2646 = 686 kJ

one mark each for left side, right side and difference **3**

Examiner's Tip
Breaking bonds uses energy and making bonds releases energy. On the left of a chemical equation all bonds are broken and on the right all bonds are made. You need to work out how many of each type of bond are broken and made, work out the energy involved and find the difference. If more energy is released than used then the reaction is exothermic, like this one.

b

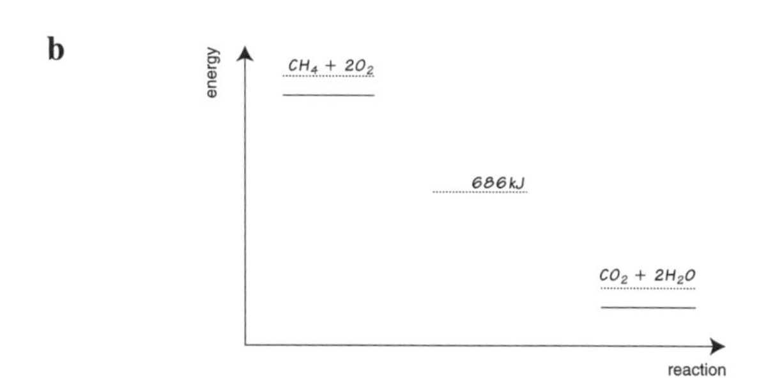

one mark for each label **3**

Examiner's Tip
Energy level diagrams show the reactants on the left and the products on the right, with the energy change involved in the reaction between them. The diagram shows what is happening in terms of energy as the reaction proceeds.

Answers: GCSE Chemistry exam 1 paper 2

Question	Answer	Mark

1 a i $CuO + 2HCl \rightarrow CuCl_2 + H_2O$
one mark for formulae, one for balance **2**

Examiner's Tip
To make copper(II) chloride by this method you would use copper(II) oxide and hydrochloric acid. From this the formulae can be worked out and the equation balanced.

ii *four from*:
Add copper(II) oxide to dilute hydrochloric acid and warm the mixture.
Filter off the excess copper(II) oxide.
Evaporate some of the water from the filtrate.
Leave the filtrate to cool.
Filter off the crystals and wash with a little distilled water.
Leave the crystals to dry. **4**
+ 1 mark for correct sequencing of stages **1**

Examiner's Tip
To make sure no unreacted acid is left, an excess of the copper(II) oxide must be used. This must then be filtered off to leave the copper(II) chloride solution. This is a dilute solution which must be concentrated before crystals will form. After filtering, the crystals will be covered with solution and must be washed before drying. The preparation involves many steps, but the question only has four marks. Any four steps will score full marks.

b i precipitation of silver chloride from a soluble silver salt with a chloride/add a solution of, e.g. sodium chloride to a solution of, e.g. silver nitrate **1**

ii filtration/decantation **1**

Examiner's Tip
If a solution of a chloride is added to a solution of a soluble silver salt like silver nitrate, the insoluble silver chloride

Question	Answer	Mark
	will form as a white precipitate. This precipitate can then be filtered off. This is the usual way to prepare insoluble salts.	
c	neutralisation	1
2 a i	Chlorine has been reduced (to chloride ions).	1
ii	Iron(II) ions have lost one electron each	1
	to form iron(III) ions.	1
iii	$2Fe^{2+} + Cl_2 \rightarrow 2Fe^{3+} + 2Cl^-$	1

Examiner's Tip
In this reaction each chlorine atom receives an electron from an iron(II) ion. Use the mnemonic OILRIG (Oxidation Is Loss and Reduction Is Gain of electrons) to work out that the chlorine atoms are reduced.
The fact that chlorine is diatomic means that the equation requires some balancing.

Question	Answer	Mark
b	Add some sodium hydroxide solution.	1
	Iron(II) forms a green precipitate.	1
	Iron(III) forms a brown precipitate.	1

Examiner's Tip
Ammonium hydroxide can be used instead of sodium hydroxide. The precipitates are iron(II) hydroxide and iron(III) hydroxide, which have quite different colours. Remember that a precipitate is a solid that appears in a clear solution during a reaction.

Question	Answer	Mark
c i	a mixture of two metals (or in this case a metal and carbon)	1
ii	Mild steel contains (a small percentage of) carbon.	1

Examiner's Tip
The usual definition of an alloy is a mixture of two or more metals, but mild steel is still called an alloy even though it is a mixture of iron with about 4% of carbon.

c iii

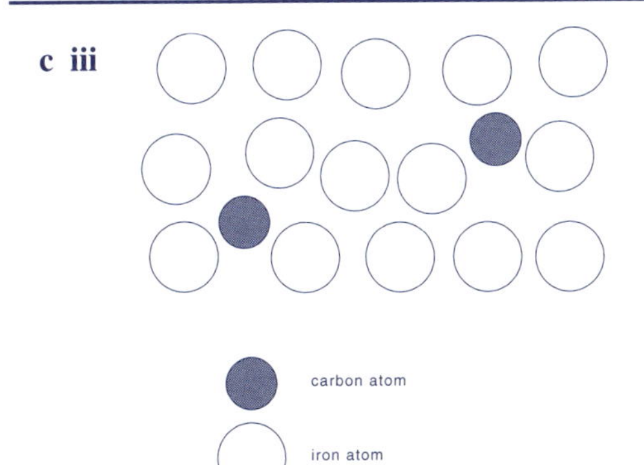

Question	Answer	Mark
	Iron and carbon atoms are different sizes/radii.	1
	The carbon atoms disrupt the regular arrangement of iron atoms.	1
	Layers of iron atoms can slide over each other.	1
	Carbon atoms prevent layers of iron atoms sliding over each other.	1
	marks can be awarded for diagram or text	

Examiner's Tip
Alloys have different properties to the pure metals because the different sizes of atoms in the alloy prevent the atoms moving around as easily. Carbon atoms are smaller than iron atoms, and therefore change the regular pattern of iron atoms into a less regular pattern.
In this question, marks can be scored from either a diagram or text. Be careful to see whether the question requires a diagram, in which case you would lose marks if you did not include one.

Question	Answer	Mark
3 a	300°C is used because reactions go faster at higher temperatures.	1
	Phosphoric acid is a catalyst to speed up the reaction.	1

Examiner's Tip
In industry, conditions for reactions are often chosen to get the product as quickly as possible. Time costs money.

Question	Answer	Mark
b i	$C_6H_{12}O_6 \rightarrow$ …2… C_2H_5OH + …$2CO_2$…..	
	one mark for each space filled in correctly	2
ii	The reaction is catalysed by enzymes.	1
	These enzymes are proteins and would be denatured at high temperature.	1
iii	The product of fermentation is a dilute solution of ethanol in water,	1
	which would need several distillations to make pure ethanol.	1

Examiner's Tip
The fermentation of sugar requires enzymes from the yeast. Since enzymes are protein molecules they are very sensitive to temperature. High temperatures would change the structure of the enzymes so that they would no longer work. The enzymes are said to be denatured. Do not describe them as killed – this would lose the mark.
Distillation of a dilute solution of ethanol produces a more concentrated solution. To make pure ethanol several distillations would be needed.

Question	Answer	Mark
c i	It is a renewable fuel.	1
ii	$C_2H_5OH \quad + \quad 3O_2 \rightarrow 2CO_2 \quad + \quad 3H_2O$	
	$24 + 6 + 16 = 46\,g$ $\quad$ $2 \times 24\,dm^3$	1
	$46\,kg$ $\quad$ $48000\,dm^3$	
	$1\,kg$ $\quad$ $48000 \times 1/46$	1
	$= 1043\,dm^3$	1

Examiner's Tip
When answering a question that gives details of mass and asks for the answer as a volume, use both in the calculation.

Question	Answer	Mark

Always begin with the equation, and use the number of molecules to work out how much mass will produce what volume of gas. Use the idea of 1 mole of a gas occupying $24 dm^3$ at room temperature and pressure. Now use the mass given in the question to work out the volume produced by this mass.
In some questions you may be asked to work out the volume at standard temperature and pressure. Under these conditions 1 mole of any gas occupies $22.4 dm^3$.
Don't worry about remembering these volumes – they will be given to you on the examination paper.

Question	Answer	Mark
4 a	wiring in computers	1
	must have a low resistance/must be a very good conductor	1

Examiner's Tip
Copper of about 95% purity is satisfactory for making water pipes, but for electrical wiring the resistance of this impure copper would be too high. The purest copper is needed in applications where wiring of very low resistance is needed, e.g. in computer circuit boards.

Question	Answer	Mark
b i	$Cu \rightarrow Cu^{2+} + 2e^-$	1
ii	$Cu^{2+} + 2e^- \rightarrow Cu$	1

Examiner's Tip
In this electrolysis the electrodes are made from copper, which is not an inert material such as platinum or carbon. The positive electrode reacts in the electrolysis to produce copper(II) ions. At the negative electrode copper(II) ions are attracted, gain electrons and are discharged as copper atoms. When you are answering a question based on an electrolysis, look to see if the electrodes are inert – and therefore do not react in the electrolysis – or reactive like copper.

Question	Answer	Mark
c	in this electrolysis the quantity of electricity passed $= 10 \times 20 \times 60 = 12000$ coulombs	1
	1 mol of copper is deposited by 2 faradays of electricity	
	64 g of copper is deposited by 193000 C	1
	$64 \times 12000/193000$ g of copper is deposited by 12000 C	
	$= 4.0$ g	1

Examiner's Tip
Remember that 1 ion of copper(II) joins with 2 electrons to make 1 atom of copper. This means that 1 mole of copper will be deposited by 2 Faradays of electricity.
A calculator gives the answer as 3.979. Since the quantities in the question were given correct to two significant figures, the answer should also be quoted to two significant figures, which is 4.0.

Question	Answer	Mark
d	Zinc is more reactive than copper.	1
	It will remain as zinc ions in solution.	1
	Platinum is less reactive than copper.	1
	It will fall to the bottom of the container as a solid.	1

Examiner's Tip
The zinc in the negative electrode will lose electrons to become zinc ions, but will remain in solution because copper is less reactive and will therefore be deposited in preference.
The platinum in the negative electrode will not react because it is less reactive than copper. It will fall to the bottom as the negative electrode gradually disintegrates. At the bottom of industrial cells of this type a number of precious metals collect, mixed with other impurities in what is called 'anode sludge'. This is collected and the precious metals are extracted from it.

Question	Answer	Mark
e	Copper is made of copper ions	1
	surrounded by a sea of electrons.	1
	The electrons move between the copper ions to conduct electricity.	1
	+ *1 mark for correct spelling, punctuation and grammar*	1

Examiner's Tip
The outer electrons of non-metal atoms are held firmly in place, but in metals these electrons move into the space between the atoms. This makes the metal atoms into ions, and means that these electrons are not held firmly. If a potential difference is applied across the metal the electrons are made to move along in the spaces between the ions.

Question	Answer	Mark
5 a i	The atoms in gases are in constant random movement.	1
	These atoms collide with each other and change direction.	1
	Since some atoms will be moving in every possible direction, the gas will move into new areas.	1
ii	Xenon has the heaviest atoms	1
	which will move more slowly than lighter atoms with the same energy.	1

Examiner's Tip
The random movement of gas particles makes the gas move into any area outside the boundaries of the vessel in which it is contained. The speed of movement of a gas particle depends on the quantity of energy it possesses. If two particles have the same energy but different mass, the heavier one will move more slowly.

Question	Answer	Mark
b i	Isotopes are atoms of the same element/atoms with the same atomic number/atoms with the same number of protons	1
	but with a different number of neutrons/a different mass number.	1

Question	Answer	Mark
ii	in 100 atoms total mass = (37 × 25) + (35 × 75)	
	= 3550	1
	so average mass of 1 atom = 3550/100	1
	= 35.5	1

Examiner's Tip
To calculate the average mass consider the mass of 100 atoms. Of these 25 will have a mass of 37 and 75 will have a mass of 35. You can use this to work out what the total mass of 100 atoms would be. Now you have to divide by 100 to get the average mass of 1 atom.

Answers: GCSE Chemistry exam 2 paper 1

Question	Answer	Mark
1 a i	copper(II) sulphate solution	1
ii	copper(II) oxide	1
iii	$CuO + H_2SO_4 \rightarrow CuSO_4 + H_2O$	
	left side	1
	right side	1
iv	neutralisation	1

Examiner's Tip
Neutralisation reactions occur not only between acids and alkalis, but also between metal oxides and acids, as in this example, and between carbonates and acids. A salt is always formed, in this case copper(II) sulphate.

Question	Answer	Mark
b	Filter or decant off the clear blue liquid.	1
	Heat the solution to evaporate off some of the water.	1
	Leave the remaining solution to cool.	1
	+ 1 mark for logical order in answer	1

Examiner's Tip
Crystals will form if a hot saturated solution is allowed to cool to room temperature. The excess copper(II) oxide must be removed first, then some of the water evaporated to form a saturated solution of the copper(II) sulphate.
To score the extra mark you must make three points in the correct order.

Question	Answer	Mark
c i	exothermic	1
ii		

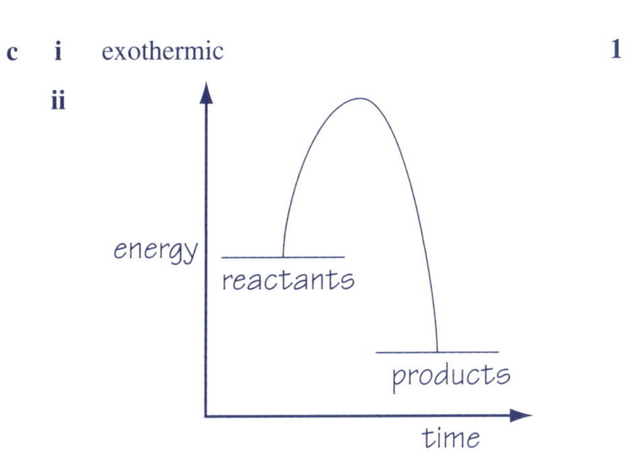

Question	Answer	Mark
	energy vs time axes drawn and labelled	1
	reactants labelled at higher energy level than products	1
	curve drawn to show progress of reaction	1

Examiner's Tip
Since an exothermic reaction gives out heat energy, the reactants must be at a higher energy level than the products. This is shown on the diagram.

Question	Answer	Mark
2 a i	water	1
ii	add to anhydrous copper(II) sulphate/cobalt chloride paper	1
	colour changes from white to blue/blue to pink	1

Examiner's Tip
The best test for water is to show that the boiling point of the liquid is 100°C, but there is not enough water in this example to do this. Anhydrous copper(II) sulphate has lost its water of crystallisation, and is white. The water restores this water of crystallisation, returning the blue colour. Always describe how to do the test, and give the colour before and after the test.

Question	Answer	Mark
b i	white precipitate/white cloudiness/lime water turns milky	2
ii	The gases contain carbon dioxide.	1

Examiner's Tip
All hydrocarbons burn to give water and carbon dioxide. Lime water forms a white solid of calcium carbonate when carbon dioxide is bubbled through it. This turns the solution cloudy – a white precipitate.

Question	Answer	Mark
c i	$2C_8H_{18} + 25O_2 \rightarrow 18H_2O + 16CO_2$	
	formulae	1
	balance	1

Question	Answer	Mark

Examiner's Tip
This is a hard equation to balance. Two molecules of octane are needed in the equation so that an even number of oxygen atoms is used. Otherwise a half molecule of oxygen would be needed. Don't be afraid of large numbers of molecules in equations – sometimes they are necessary.

ii When the octane does not have sufficient oxygen for complete combustion **1**
some of the carbon in the octane does not combine with oxygen. **1**

Examiner's Tip
Hydrocarbons only burn completely to water and carbon dioxide if there is plenty of oxygen available. In air there is not enough oxygen, so the octane does not burn completely. All of the hydrogen forms water, but some of the carbon will form carbon monoxide or carbon. The carbon gives a sooty deposit.

3 a

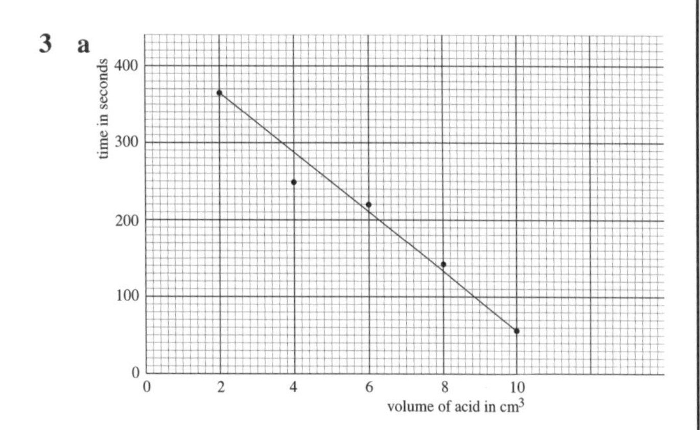

axes correctly drawn and labelled, including units **1**
all points plotted to + or − half a square **1**
a best fit line drawn ignoring the second point **1**

Examiner's Tip
Axes need to be sensibly scaled and labelled with the thing being plotted, e.g. volume of acid, and the units, e.g. cm^3. There is just one mark for doing this correctly for both axes! Plotting of the points must be accurate. Mark them clearly with a circle or cross. The best fit line must ignore any anomalous results.

b i This kept the total volume the same in each case, **1**
otherwise the concentration would not have been proportional to the volume of acid added. **1**

ii temperature (of the acid and water mixture)/stirring **1**

Question	Answer	Mark

Examiner's Tip
By using different volumes of acid diluted with water to the same total volume each time, Sarah made sure that the concentration was proportional to the volume of acid added. This could then be plotted to give the graph. Since the rate of a reaction increases with increase in temperature, this has to be kept constant if the investigation of rate with concentration of acid is to be a fair test.

c i 4 cm^3 of acid **1**

ii incorrect measurement of volumes or time/not constant temperature/inconsistent stirring/inconsistent tablets **1**

Examiner's Tip
The result for 4 cm^3 obviously does not fit onto a straight line, which this graph should have. There are many possible reasons for this, and any sensible suggestion would score the mark in (ii).

d i Rate increases with increase in concentration. **1**
Rate is directly proportional to concentration. **1**

ii In order to react the acid particles need to collide with the solid sodium carbonate in the tablet. **1**
At higher concentration there are more particles of acid per cm^3, **1**
therefore more particles collide with the sodium carbonate each second. **1**

Examiner's Tip
As in many questions, the number of marks indicated for each part must be carefully noted. In both (i) and (ii) it would be easy to write less than the number of points needed for full marks. The rate of a reaction depends on the number of particles which collide each second. Of these collisions a proportion will result in the formation of products. The same proportion of a larger number of collisions will result in the formation of more product in a certain length of time, i.e. a greater rate of reaction.

4 a i mass of copper = 15.52 − 12.64 = 2.88 g **1**

ii mass of oxygen = 15.88 − 15.52 = 0.36 g **1**

b moles of copper = $\frac{2.88}{64}$ = 0.045 **1**

moles of oxygen = $\frac{0.36}{16}$ = 0.0225 **1**

mole ration of copper to oxygen = 0.045:0.0225 = 2:1 **1**

formula must be Cu_2O **1**

Question	Answer	Mark

Examiner's Tip
The masses of copper and oxygen are easily worked out by subtraction, but be careful that you are subtracting the correct figures! The mass of each element in the copper oxide is used to work out the number of moles by dividing mass by atomic mass. The number of moles of each element then gives the mole ratio. The tricky bit is converting this to a whole number ratio. Simply divide the larger value by the smaller one. The whole number ratio gives the number of atoms of each element: two for copper and one for oxygen.

c	Sodium is much more reactive than copper,	**1**
	so it is not so easy to remove the oxygen from sodium oxide/ hydrogen will not remove the oxygen from sodium oxide.	**1**

Examiner's Tip
The more reactive a metal, the more energy is needed to pull oxygen away from the metal in the oxide. Hydrogen is not a strong enough reducing agent to remove the oxygen from sodium oxide.

5 a	**i**	catalytic	**1**
		cracking	**1**

Examiner's Tip
It would be very easy to answer 'cracking' for this question and get only one mark. Do not forget to look at the number of marks to be given for the answer and make sure you write a separate point for each one.

	ii	catalyst	**1**
	iii	small pieces have greater surface area	**1**
		giving more collisions with the hydrocarbon per second	**1**

Examiner's Tip
The aluminium oxide is a solid catalyst for a reaction in which a gas is reacting. The amount of contact between the catalyst and the gas will have a large effect on how fast the reaction will go. Don't forget to use the ideas of time when you are writing about rate of reaction.

b	**i**	ethene	**1**
	ii	add bromine water	**1**
		changes from red/orange to colourless	**1**

Examiner's Tip
The gas must be an alkene, and ethene is the alkene with two carbon atoms. Bromine water is the standard test for an alkene. The bromine reacts with the alkene and therefore is used up, leaving no colour in the mixture. An alkane will not decolourise bromine water.

c

```
  H H H H H H H H H H
  | | | | | | | | | |
H-C-C-C-C-C-C-C-C-C-C-H →
  | | | | | | | | | |
  H H H H H H H H H H

      H H H H H H H H        H     H
      | | | | | | | |         \   /
    H-C-C-C-C-C-C-C-C-H  +     C=C
      | | | | | | | |         /   \
      H H H H H H H H        H     H
```

one mark for each graphical formula **3**

Examiner's Tip
Remember to show each atom and each bond in a graphical formula, and don't forget the double bond in ethene. It is easy to give the wrong number of carbon atoms when there are so many, so count them to make sure they are correct for decane and octane.

6 a	**i**	sandstone A	**1**

Examiner's Tip
Youngest rocks are always at the top of a sequence.

	ii	basalt	**1**

Examiner's Tip
Quite a tough question – the two igneous rocks that you have to know about for GCSE are basalt and granite. Questions are often about granite, shown as an igneous intrusion, but this is about basalt as a lava flow.

	iii	limestone	**1**

Examiner's Tip
Learn that limestone, marble and chalk are all mostly calcium carbonate.

b	Since it contains crystals it is likely to be an igneous rock/formed from lava solidifying; since it contains very small crystals it must have cooled quickly/above ground.	**2**

Examiner's Tip
If you see two marks in a question like this, don't just put 'quick cooling' – think what else you could say.

Question	Answer	Mark
c	Fault X–Y has affected the mudstone layer but fault A–B has not; since the younger layers are on top of the older layers this must mean that X–Y occurred after A–B.	2

Examiner's Tip
In this sort of question, think about the order in which events must have taken place over geological time. Which layer was deposited first? Then what was next? When did fault A–B occur? You will then realise that X–Y must have taken place later than fault A–B.

Question	Answer	Mark
7 a i	respiration	1
ii	$6CO_2 + 6H_2O$ (+ energy from sunlight) $\rightarrow$ $C_6H_{12}O_6 + 6O_2$	
	formulae	1
	balance	1
iii	from the sun/sunlight	2

Examiner's Tip
It is easy to get respiration and photosynthesis mixed up on the carbon cycle. Remember that respiration happens in both plants and animals, and produces CO_2, but photosynthesis happens in plants only, and uses up CO_2. You will get both marks even if the energy is not included in the equation in (ii). In (iii) you will not get the mark for 'light'.

Question	Answer	Mark
b i	As carbon dioxide was removed from the air by photosynthesis,	1
	an equal amount	1
	was returned to the air by respiration.	1

Examiner's Tip
The important idea here is that the removal and return of carbon dioxide was balanced, so that the percentage in the air remained constant.

Question	Answer	Mark
ii	carbon dioxide percentage has increased	1
	caused by an increase in the burning of fossil fuels/ destruction of rain forests	1
c	Hydrogen and helium have low densities.	1
	They escaped from the Earth's atmosphere.	1
	Ammonia reacted with oxygen to produce nitrogen and water	1
	Bacteria (nitrifying or dinitrifying) removed ammonia.	1
	+ 1 mark for correct spelling, punctuation and grammar	1
	(Choose the best sentence and look for capital letters, spelling and grammatical errors.)	

Question	Answer	Mark
8 a	reversible/can go both ways/can form an equilibrium	1

Examiner's Tip
Any of the above answers are acceptable although saying that it shows that the reaction is reversible is probably the easiest to remember.

Question	Answer	Mark
b	nitrogen from air	1
	hydrogen from crude oil/natural gas	1
c	Increasing the pressure increases the rate of reaction (because there are a greater number of successful collisions between the greater numbers of particles present); increasing the pressure increases the yield (because it pushes the equilibrium to the right and produces more ammonia).	2

Examiner's Tip
This is a difficult A* question. You have to use your knowledge of reaction rates to realise that increasing the pressure of gases increases their concentration and so will mean that more successful collisions will take place because there are more particles present. You also have to understand that since this equilibrium equation shows fewer molecules on the right hand side of the equation, increasing the pressure will force the equilibrium to the side with the smaller number of molecules, in this case to the right. This is often referred to as 'Le Chatelier's Principle' but knowing this name is not part of every examination board's specification.

Question	Answer	Mark
d	*Three from:* Although the yield is high at low temperatures the rate of reaction is slow; this is because gas particles have less energy and so there are fewer successful collisions; using a higher temperature will give a lower equilibrium yield but this lower yield will be obtained much more quickly; it is important economically to produce ammonia quickly and so higher temperatures are used which give a fast reaction rate; a catalyst can be used to help speed up the rate at which the equilibrium is established.	3

Examiner's Tip
Another difficult question but the mark scheme enables you to score three marks without giving every single answer on the list. The examiner will look to see that you have understood the idea that too low a temperature will give too slow a rate of reaction because of fewer successful collisions between particles.

Question	Answer	Mark

e Ammonia is used to make artificial fertilisers; **1**
this was important because of the rapidly expanding world population requiring increased food production. **1**

f $2 \times (14 + 3)$ **1**
$= 34$ tonnes **1**

9 a atomic number increases from Na – Cl – Ar/increases 11 – 17 – 18 **1**
number of electrons in outer shell increases: Na – 1, Cl – 7 and Ar – 8 **1**

Examiner's Tip
It is important to write about the number of protons, since this decides the element's position in the Periodic Table, and the number of electrons in the outer shell, since this determines the chemical behaviour of the element.

b Sodium is a very reactive metal. **1**
Chlorine is a very reactive non-metal. **1**
Argon is an unreactive gas. **1**

c Sodium has one electron in its outer shell which is easily lost to get the stable electronic structure of argon – a typical metal property. **1**
Chlorine has seven electrons in its outer shell and easily gains one more to get the stable electronic structure of argon – a typical non-metal property. **1**

Question	Answer	Mark

Examiner's Tip
These two questions look at the relationship between the number of electrons in the outer shell of an atom and its chemical properties. The fact that chemical bonding leads to each atom having a full outer electron shell, which is the same electronic structure as a noble gas, is an essential feature of chemistry.

10 a *Either*
Calcium sulphate has a low solubility in water **1**
Calcium sulphate is precipitated **1**
or
Sea creatures with shells (crustaceans) use calcium ions to build up shells **1**
Shells are calcium carbonate **1**

b *Either*
$Ca^{2+}(aq) + SO_4^{2-}(aq) \rightarrow CaSO_4(s)$ **2**
One mark for left hand side and one mark for right hand side.
or
$Ca^{2+}(aq) + CO_3^{2-}(aq) \rightarrow CaCO_3(s)$ **2**
One mark for left hand side and one mark for right hand side.

Examiner's tip
New specifications have questions on compositions of the oceans. There are two correct answers to this question. You will not be penalised if you miss out state symbols.
The table helps you with the correct formulae of calcium and sulphate ions.

Answers: GCSE Chemistry exam 2 paper 2

Question	Answer	Mark

1 a

name	formula	molecular mass	boiling point in°C	
methane	CH_4	16	–161	
ethane	C_2H_6	30	........–88°C*....	1
..propane..	C_3H_8	44	–42	1
butane	....C_4H_{10}...	58	–1	1
pentane	C_5H_{12}	72	36	

** (any negative value between –70 and –130 accepted)*

Examiner's Tip
Look carefully at the information in the table and use it to help fill in the blanks. You will see that each set of information has a pattern. Use the pattern to work out the correct value for the blank box.

b A homologous series is a series of compounds each differing from the last by the same group of atoms or each having the same general formula. **1**
Each alkane has CH_2 more than the one before
or
the general formula C_nH_{2n+2}. **1**

Examiner's Tip
The question says 'as it applies to the alkanes' so your answer must say exactly how the term does apply to the alkanes.

Question	Answer	Mark

c i Structural isomers have the same molecular formula **1**
but different structural (displayed) formulae. **1**

ii

```
    H H H H              H H H
    | | | |              | | |
  H-C-C-C-C-H          H-C-C-C-H
    | | | |              | | |
    H H H H              H | H
                           C
                         H | H
                           H
```

one mark for each diagram **2**

Examiner's Tip
It is easy to draw a straight chain alkane with one of the carbon atoms pointing up or down and think this is an isomer. To be different the structural formula must actually have a carbon atom joined on in a different place.

2 a

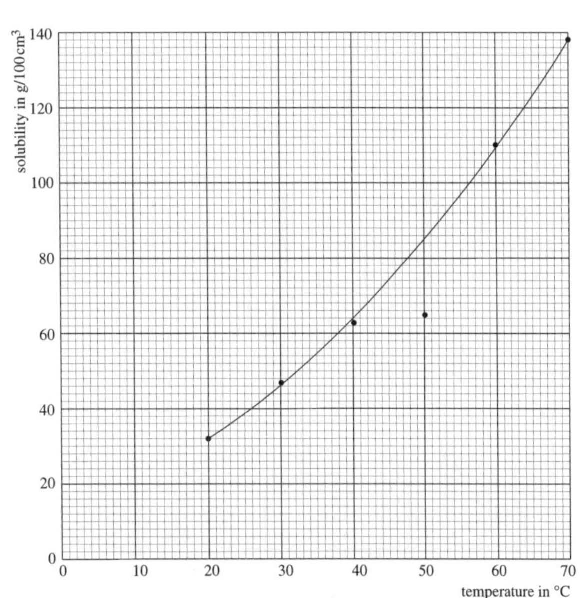

sensible axes, each correctly labelled with name and unit **1**
all points plotted correctly **1**
line passing through all points except the one for 50°C **1**

Examiner's Tip
The value for 50°C is obviously incorrect and should be ignored when drawing the best fit line.

b Solubility increases with increase in temperature **1**
but is not directly proportional/increase becomes greater with higher temperature. **1**

Question	Answer	Mark

Examiner's Tip
You would not get a mark for saying 'solubility increases with temperature'. It increases with *increase* in temperature.

c i 50°C **1**

ii 86 g per 100 cm^3 water (+ or − 2 g) **1**

Examiner's Tip
You can read off the correct value for 50°C on your graph. There is some allowance for your curve not being exactly the same shape as the Examiner's.

d i Crystals would appear **1**
and grow bigger. **1**

ii The solution at 70°C would be saturated with potassium nitrate. **1**
This would come out of solution as it cooled, so making crystals 'grow'. **1**

Examiner's Tip
Note that there are two marks for each of these answers. Always look carefully at the number of marks and make sure you write one point for each mark.
For example, writing just 'crystals appear' for (i) would only score one mark.

3 a i no more bubbles/effervescence/hydrogen given off **1**

ii Zinc is higher in the reactivity series/ more reactive. **1**

b i $HCl + NaOH \rightarrow NaCl + H_2O$
mole ratio is 1 mole NaOH reacting with 1 mole HCl **1**
so 14.6 cm^3 0.5 mol/dm^3 NaOH reacts with 14.6 cm^3 0.5 mol/dm^3 HCl **1**

Examiner's Tip
It is important to write the equation between hydrochloric acid and sodium hydroxide to find out that the mole ratio is 1:1. If you simply assume this you will lose a mark. Since the two solutions have the same concentration they will react in equal volumes.

ii volume of 0.5 mol/dm^3 HCl reacting with zinc = 25 − 14.6 = 10.4 cm^3 **1**

Examiner's Tip
This is a simple subtraction of the volume of acid used from the volume used originally to react with the zinc in the coin.

Question	Answer	Mark
iii	mole ratio from equation is 1 mole Zn to 2 moles HCl 10.4 cm³ 0.5 mol/dm³ HCl contains $0.5 \times \frac{10.4}{1000} = 0.0052$ moles	1
	moles Zn reacted = 0.5×0.0052 = 0.0026 moles	1
	mass Zn reacted = $0.0026 \times 65 = 0.169$ g	1

Examiner's Tip
You need to look back to the equation to see what the mole ratio of zinc to hydrochloric acid is. It is 1:2, so if you used 1:1 by mistake you would lose one mark. The volume of HCl can be used to calculate the moles of HCl, which then must be halved to get the moles of Zn.
Finally the moles of Zn must be multiplied by the relative atomic mass of zinc to get the mass in g.

Question	Answer	Mark
iv	% zinc in coins = $100 \times \frac{0.169}{0.5}$	1
	= 33.8 %	1

Examiner's Tip
The final stage is simply to divide the mass of zinc by the mass of the coin and multiply by 100 to get the %. Many candidates forget to multiply by 100.

Question	Answer	Mark
4 a i	stops rusting/stops corrosion/looks more attractive	1

Examiner's Tip
Most electroplating is carried out either to protect the original from corrosion or to improve the appearance of the product – often to make it look more expensive.

Question	Answer	Mark
ii	metal M/carbon/platinum	1
iii	the anode needs to be of the same metal as the electroplating to replace the ions removed from solution/the anode needs to be inert	1

Examiner's Tip
The answer to (iii) must match your answer to (ii). If a metal was used for the anode which then reacted in the electrolysis, e.g. copper, this would put ions of this second metal into the solution and spoil the electroplating.

Question	Answer	Mark
b i	mass of M = 10.94 − 10.50 = 0.44 g	1

Examiner's Tip
This is simply a subtraction of the mass before electroplating from the mass after electroplating.

Question	Answer	Mark
ii	quantity of electricity = $2 \times 12 \times 60$	1
	= 1440 coulombs	1

Examiner's Tip
The amount of electricity used is calculated using the formula Q = It, in other words: coulombs = amps × minutes × 60. Note that time is in seconds for this calculation, hence the × 60. The answer is in coulombs of electricity.

Question	Answer	Mark
iii	1 mole of M would need $1440 \times \frac{59}{0.44} = 193091$ coulombs;	1
	this is 2 Faradays, so the number of positive charges on an ion of M = 2.	1

Examiner's Tip
The relative atomic mass of M is given in the question as 59. The answers from parts (i) and (ii) can be used to calculate the number of coulombs needed to deposit 1 mole of M. Since this value is 2 × 96500, i.e. 2 Faradays, the charge on the ion of M must be 2+. You need to set out your answer carefully, showing each calculation.

5 a

element	protons	arrangement of electrons	Group in the Periodic Table	metal or non-metal	Mark
Q	8	2, 6	6	non-metal	1
X	11	2, 8, 1	1	metal	1
Y	17	2, 8, 7	7	non-metal	1
Z	18	2, 8, 8	0	non-metal	1

Examiner's Tip
These answers can easily be worked out by using information already in the table, e.g. X has the electronic arrangement 2, 8, 1; this adds up to 11 electrons, so the atom must also have 11 protons.

Question	Answer	Mark
b	Y	1

Examiner's Tip
A crystalline salt can be formed from a metal atom from Group 1 and a non-metal atom from Group 7 joined together to make an ionic compound. X is a metal in Group 1 and Y is a non-metal in Group 7.

Question	Answer	Mark
c	X_2Q	1

Examiner's Tip
An atom of X has one electron in its outer shell to donate, and an atom of Q needs two electrons to join its outer shell for it to be full with 8. Two atoms of X will give one electron each to one atom of Q to form an ionic compound.

Question	Answer	Mark

The formula must therefore contain two atoms of X and one atom of Q.

d

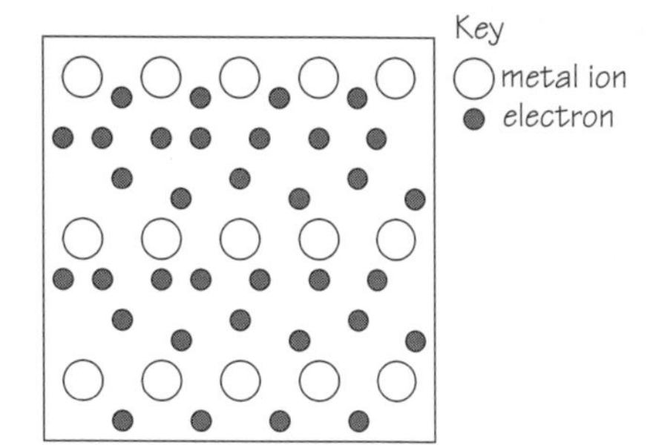

diagram similar to above **1**
electrons between metal ions are mobile **1**
and flow when pd is applied to give current through metal **1**

Examiner's Tip
The diagram should clearly show metal ions with a 'sea' of electrons between them. It is a good idea to put a key next to the diagram. The 'sea' of electrons is responsible for the conduction of electricity through the metal. When a potential difference is applied these electrons move though the metal.

e Element Y is made of molecules **1**
each containing two atoms. **1**

6 a i Calcium hydrogen carbonate decomposes on heating. **1**
Calcium carbonate is deposited as a solid. **1**
This removes the calcium ions causing hardness from solution. **1**
+ 1 mark for correct use of scientific language **1**

ii The water from source A must contain permanent hardness
(as well as temporary hardness). **1**

Question	Answer	Mark

Examiner's Tip
Temporary water hardness is caused by calcium hydrogen carbonate dissolved in the water. When this is decomposed by heat it forms insoluble calcium carbonate which is precipitated as a white solid. Since the calcium ions responsible for the water hardness are now in the solid calcium carbonate, not in solution, the water is soft. Permanent water hardness is caused by calcium sulphate, which is not decomposed by heat and therefore causes the water still to be hard after boiling.

b i A precipitate of calcium carbonate is formed **1**
removing the calcium ions from solution **1**
$Ca^{2+}(aq) + CO_3^{2-}(aq) \rightarrow CaCO_3(s)$ **1**
+ 1 mark for correct spelling, punctuation and grammar. **1**

ii All calcium ions are removed when sodium carbonate is used. **1**

Examiner's Tip
Since calcium carbonate is insoluble in water, adding carbonate ions, in the sodium carbonate, to the hard water will cause calcium carbonate to form a precipitate. Without calcium ions the water is soft. It does not matter whether the calcium ions are from permanent or temporary hardness, they will all be removed from solution.

c scum formed with soap/kettles fur up **1**

Examiner's Tip
Calcium ions form an insoluble precipitate, called scum, with soap. This means that more soap has to be used, and the scum can stain clothing. If temporary hardness is present in water boiled in a kettle, the calcium carbonate formed collects in the kettle; this is called furring up.
It can cause failure of the element in an electric kettle or washing machine.

Examiner's report and grade predictor

WHAT IS IN THE EXAM?

The first examinations of new specifications (or syllabuses) were held in 2003.

For all Chemistry GCSE courses candidates had to complete two papers – one in the Core material (common with the Chemistry in Double Award Science) and one which included extra extension material added to make up the Chemistry specification. Each Awarding Body (or Examining Board) added different extension material. The extension material is not more difficult but gives each specification its identity.

The marks in these two papers added together make up eighty per cent of the marks, with twenty per cent coming from the Coursework (Sc1). The grade is then calculated from the total mark achieved.

From 1998 questions on the Higher tier have tested the whole specification. Some statements are identified in the specification as only to be examined on Higher tier but the examinations test the whole specification not just these statements. From 2003, questions could not be set on material you covered in Key Stage 3.

Most people think that GCSE examinations concentrate on testing the knowledge you have learnt. This was certainly the case twenty or thirty years ago. Then examinations concentrated on testing recall of knowledge and there was much more factual information on the syllabus. This is why people keep saying that the examinations are getting easier. You now have to recall fewer facts but you are required to do much more with the knowledge you have.

Ignoring the Coursework, of the remaining eighty per cent of the marks, just sixty per cent is allocated to knowledge and understanding. Of this, one third is allocated to recall of facts and the remainder to showing an understanding of the factual knowledge. Being able to recall the formula of sulphuric acid as H_2SO_4 is recall but being able to write a chemical equation for a reaction involving sulphuric acid is understanding.

The final fifteen per cent is allocated to higher level skills of interpretation, evaluation, etc. This could involve using your knowledge from different parts of the specification and applying it to a different and new situation. It is here that candidates find the most difficulty.

Examination papers have to have questions requiring longer answers. These are called Continuous and Extended Writing. For Higher tier about 15 marks have to require answers of two sentences and 10 marks have to require longer answers. It is common on candidates' papers to see that performance on these questions is disappointing compared with the rest of the papers. Where candidates often go wrong is to fail to give complete answers and to get the answer out of the correct order.

If you are having difficulty in Chemistry and your aim realistically is to achieve a Grade C, you are more likely to achieve this by taking the Foundation tier papers.

NEW FEATURES IN 2003

Quality of Written Communication (QWC)

About four marks on each paper have to be allocated to QWC. On these questions – and you will be told which questions they are – marks can be awarded by the examiner for your ability to either:

- write in sentences
- use correct spelling, punctuation and grammar
- use correct scientific terms.

Many candidates write answers in bullet points rather than sentences. You should make sure you use sentences in any of these questions.

Ideas and Evidence questions

Five per cent of the questions on the paper will be based upon Ideas and Evidence. You will probably not know which questions they are. You are expected to use the information to give your opinions but you should always support your opinions with scientific information either from the paper or from your knowledge.

Ideas and Evidence can be divided into four sections.

1. How scientific ideas are presented, evaluated and disseminated.
2. How scientific controversies can arise from different ways of interpreting empirical evidence.
3. Ways scientific work may be affected by the context in which it takes place.
4. Ways to consider power and limitations of Science in addressing industrial, social and environmental questions.

HOW TO IMPROVE YOUR GRADE

If you are taking GCSE Chemistry, your grade will depend upon your performance in the two written papers and Coursework.

Frequently candidates who do well on the Core Chemistry paper do less well on the Extension paper. There can be many reasons for this. Possibly less time has been allocated for this extension material both in school or college but also in revision.

An examination cannot cover everything in the specification but it does try to sample it all. Do not miss out sections in your revision. Concentrate on the whole specification and pay special attention to learning definitions and the correct use of scientific terminology.

WHAT IS NEEDED FOR A GRADE A?

Contrary to many people's beliefs, grade A is not determined each year by awarding the grade to a fixed percentage of candidates or by awarding a grade A to those candidates who achieve a fixed mark. It is done by inspection of the papers and awarding grade A to those candidates who meet the criteria that have been agreed nationally for grade A.

A grade A candidate should be able to:

1. use detailed knowledge and understanding to devise a strategy for a task;
2. identify key factors in the task and control conditions;
3. make predictions;
4. present data appropriately and use knowledge from different sources;
5. recognise and explain anomalous results;
6. draw appropriate graphs choosing suitable axes;
7. use scientific knowledge and understanding to draw conclusions;
8. identify shortcomings in evidence;

9. use a range of apparatus with the correct precision and skill:

10. make precise measurements and systematic observations;

11. select which observations and measurements are relevant;

12. recall information from all areas of the specification;

13. use detailed scientific knowledge and understanding in a range of applications;

14. detect patterns and draw conclusions when information comes from different sources;

15. draw information together and communicate knowledge effectively;

16. use scientific or mathematical conventions to support arguments;

17. use a wide range of scientific and technical vocabulary.

Some of these criteria can be met in Coursework (Scl) but most can also be demonstrated in written papers.

WHAT EXTRA IS REQUIRED FOR AN A* GRADE?

Having established what is required to be awarded a Grade A you might be interested to know what is required for an A* grade. There are at present no A* criteria.

When the Awarding committee is awarding grades it is asked to fix marks for Grade A and Grade C. This is done paper by paper. Suppose on a particular paper the Grade A mark was fixed at 70 and the grade C mark at 50. (These numbers have been chosen only to keep the arithmetic that follows simple). The Grade B mark is then fixed arithmetically half way between 50 and 70, i.e. 60. The Grade A* is then fixed the same number of marks above A as B is below it. In the example we have used A* would be fixed at 80. At this point it would be customary to look at papers around this mark to confirm that they were worthy of A*. What does this tell you? Grade A* is a very high standard and relatively few are awarded.

As there are no criteria it is not as clear what examiners are looking for as it is at Grade A. The following points may help you.

- Generally as the Grade A* boundary is a high mark, there is no scope for a bad answer to any question on the paper. A grade A* candidate scores well on questions.
- A grade A* candidate uses scientific language routinely and confidently. It is worthwhile working through a glossary of scientific terms or a scientific dictionary to clarify the exact meaning of all terms and then trying to use them correctly.
- A grade A* candidate brings information from different parts of the specification together in an answer.
- Grade A* candidates in Chemistry have a clear idea of models of atomic and molecular structure and of the kinetic model for particles in solids, liquids and gases. They can use these ideas of models to explain ideas such as diffusion, rate of reaction, reactivity, bonding between atoms and rates of reaction.
- Grade A* candidates in Chemistry can write balanced symbol equations and ionic equations, using state symbols if requested. They also have a clear understanding of the Periodic Table, and can use ideas from the table to predict the behaviour of elements and compounds.
- Chemistry calculations are often regarded as a very good indicator of a high level candidate. Grade A* candidates should be able to handle a full range of calculations, including those using equations, finding the formula from data, simple titration calculations and percentage calculations. They set out calculations clearly, showing each step in the process. A grade A* candidate gives answers to the correct number of significant places and with correct units.
- While many good candidates can substitute numbers into equations and calculate answers, grade A* candidates understand mathematical relationships. Most calculations in Chemistry involve ideas of ratio or proportion. A grade A* candidate fully understands these ideas and can apply them to any situation.

HOW TO ASSESS YOUR GRADE

The matrix below suggests grades that you might have expected to achieve with different scores on *each* exam. It is an indication only and does not imply that this is the grade you will receive in the real examination.

A*	135–160
A	110–134
B	90–109
C	70–89
D	45–69